MathBridge

A Parent's Guide to Building Math Confidence, Reducing Anxiety, and Supporting Lasting Learning

By

Lawrence Mayaki

iS Mayaki
CONSULTING & COACHING

First Edition

ISBN: 979-8-90599-863-8

Publisher: Staten House

Disclaimer

This book draws upon contemporary research in mathematics education, cognitive science, educational psychology, and child development, together with the author's extensive professional experience working with primary school-aged children and their families. It provides educational guidance for parents supporting children with math anxiety and low confidence. It is not intended to diagnose, treat, or replace professional psychological, therapeutic, or medical support.

Privacy and Composite Examples

The examples and case studies in this book are based on the author's professional experiences working with children and families over many years. To protect privacy, identifying details have been changed, and some examples represent composite experiences drawn from multiple individuals.

Preface

Across more than two decades in education, my work underwent a profound shift twelve years ago. I began intentionally studying math anxiety and low math confidence. Again and again, I saw that many children were not struggling in mathematics because they lacked cognitive ability. Instead, they were struggling because mathematics had become entangled with fear, embarrassment, performance pressure, or the internalized belief that they simply were "not good at math."

At first, my focus was primarily on instructional and teaching strategies. Within a school settings, we call it math support. Over time, however, I began thinking much more about the emotional experiences children bring into mathematics classrooms and home learning sessions. I started asking difficult, systemic questions:

- Why do some highly capable children shut down entirely during mathematics?
- Why do others panic before they have attempted a problem?
- What role do confidence, anxiety, relationships, and emotional safety play in learning mathematics?
- And how can parents and teachers work together more effectively to support every child?

Those questions led me into years of reading, observation, research, reflection, and direct consultation with families. All of this learning led me to one important realization. Both children and adults often carry far more anxiety than either of them knows how to put into words.

Over time, I developed a deeper understanding of math anxiety, confidence, memory, emotional regulation, and the need for parental support. I also began sharing practical and actionable insights with parents whose children were struggling with mathematics.

Over the years, I have watched remarkable transformations happen. I have seen children who once deployed endless avoidance tactics begin asking for extra math problems to solve at home. I have heard parents tell me, with immense relief, "My child now says they actually enjoy math." I have worked face-to-face with children who moved out of fear and chronic shutdown, moving toward a higher confidence level, persistence, and independence.

In one memorable case, a student who had previously struggled significantly with confidence and anxiety later performed exceptionally well in school mathematics competitions after gradually rebuilding both their mathematical understanding and self-belief. What mattered most was not the accolades or the competition itself. What mattered was watching a child who once doubted their own ability begin to trust their own thinking again.

I have also experienced the joy of former students returning to me year after year from middle school to share how much more confident they felt approaching mathematics compared to their early days. Those moments reinforced a vital truth for me: confidence changes outcomes. When children believe they can think, recover, and improve, their entire relationship with learning shifts.

This book was written because I believe children deserve emotionally safe mathematical experiences both at school and at home. I also believe parents deserve practical, research-informed guidance that empowers them to support their child's learning without unintentionally increasing fear, urgency, or performance pressure.

My hope is that MathBridge helps parents see far beyond marks, speed, and worksheets, and begin noticing the real emotional experiences children carry into the learning space. I hope it helps families create calmer conversations, healthier learning relationships, and everyday moments where children begin to trust their own thinking again.

Most importantly, I hope children begin to experience mathematics not as a source of recurring fear or self-doubt, but as a space where mistakes are survivable, independent thinking is supported, and growth remains entirely possible.

MathBridge is not about creating perfect mathematicians. It is about helping children build confidence, resilience, flexible thinking, and a healthier relationship with challenge itself—not only in mathematics, but in learning and life far beyond it.

Lawrence Mayaki
iSMayaki Consulting & Coaching

Preface

How to Use This Book

Part One: Foundational Guide

The Parent Support Journey

How to Use This Book

MathBridge is designed differently from a traditional workbook. It combines a Foundational Guide for parents with a practical Parent Support Journey.

Part One: The Foundational Guide

The first section of this book explores the emotional and cognitive factors that influence math confidence. You will learn how anxiety affects learning, why some children become trapped in cycles of avoidance and self-doubt, and how parents can create conditions that support confidence, resilience, and long-term mathematical growth.

Part Two: The Parent Support Journey

Once you have completed the foundational guide, the book transitions into the Parent Support Journey, designed to rebuild your child's relationship with mathematics. Through carefully sequenced learning engagements, you will support the development of confidence, numeracy, mathematical thinking, problem-solving skills, and increasing independence while reducing the anxiety that often prevents children from demonstrating their true capabilities.

Each week follows the same structure:

Weekly Purpose

The key focus and intended outcomes for the week, helping you understand both the emotional and mathematical goals of the learning experiences.

Parent Learning Focus

Practical guidance to help you adjust your approach, strengthen productive learning habits, and better support your child during home learning.

The Science Behind This

Research-informed explanations that help you understand why the strategies work and how they support learning, confidence, and long-term retention.

Companion Resources

Companion resources designed to support each week's learning, available through the MathBridge™ Companion Portal.

The Companion Portal

To keep the book focused on guidance and implementation, worksheets, games, templates, and printable resources are provided through the MathBridge™ Companion Portal.

At the end of each weekly section, you will find QR code that provide access to the resources you'll need for each week. By separating the guidance from the printable materials, you can first understand the emotional and cognitive needs behind your child's struggles and then select the resources that best support their learning.

You do not need to complete every learning engagement perfectly, nor should you feel pressured to move through the journey at a particular pace. Every child develops confidence, understanding, and independence differently. Use the ideas, strategies, and resources in a way that best supports your child's needs, while remembering that meaningful progress is often built through small, consistent steps over time.

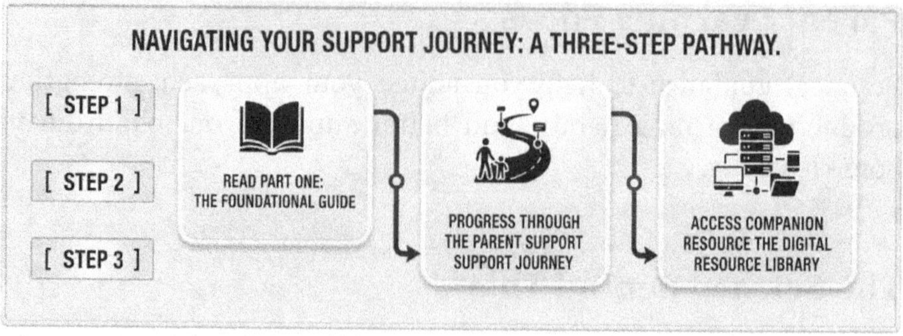

A Final Note Before We Begin

The most effective way to use this book is to read the foundational guide before beginning the Parent Support Journey. Doing so will help you understand the emotional, cognitive, and behavioral factors that influence your child's relationship with mathematics and will provide the context you'll need to apply the weekly strategies effectively.

As you move through the Parent Support Journey, try to focus on steady progress rather than perfection. The goal is not to complete the learning engagement activities as quickly as possible, but to create an environment that reduces anxiety, strengthens confidence, supports conceptual understanding, and encourages increasing independence over time.

Every child progresses differently. Use the ideas, routines, and resources in a way that best supports your child's unique needs.

Chapter 1: Introduction

Most evenings, in homes across the world, a quiet and sometimes stressful ritual takes place. A workbook is opened, a pencil is sharpened, and a child sits down to face their mathematics home learning. Within minutes, the atmosphere in the room shifts. The child might begin to stall, find sudden distractions, or break down into tears of frustration. On the other side of the table, a well-meaning parent feels a knot tighten in their stomach—torn between the desire to help and an overwhelming sense of helplessness.

If this scene feels familiar, you are not alone, and your child is not broken.

What looks like a battle over fractions, multiplication, or long division is rarely just an academic struggle. More often than not, it is an emotional hurdle. Mathematics can trigger feelings of vulnerability and anxiety in ways that many other subjects do not. Unlike subjects where a student can write an approximation or express an opinion, elementary mathematics feels unyielding to a child. An answer is either entirely right, or it is completely wrong. This binary nature can make every mistake feel like a definitive judgment on their cognitive ability.

Many children do not fear mathematics because they are incapable of learning mathematics. They fear mathematics because it has become connected to confusion, embarrassment, comparison, mistakes, emotional tension, or the internalized belief that they are simply "not good at math."

For parents, this can be incredibly difficult to watch. A child who once counted happily, played naturally with numbers, or enjoyed simple logic games may suddenly begin to avoid home learning, rush through tasks, seems careless, or simply say "I hate math" before the work even begins. These moments can easily feel like defiance, but they usually reflect something much deeper about confidence, emotional steadiness, and the way a child now experiences learning mathematics.

MathBridge was created to help parents understand and navigate that story. This is not a book about pushing children harder. It is a guide for helping children feel safe enough to think, supported enough to try, and confident enough to recover when mathematics becomes genuinely difficult.

This book focuses specifically on children in Grades 3–5 (typically ages 8–11) because this is often the stage when mathematics begins to feel significantly more demanding. Children move rapidly from early number work into multi-step reasoning, place value, regrouping, fractions, word problems, multiplication, division, and complex mathematical explanations. At this exact same time, they are forming their core beliefs about themselves as learners.

The strongest home support does not attempt to replace school teaching. Instead, it strengthens the emotional and learning conditions that allow school learning to happen. A parent does not need to become a mathematics expert to make a meaningful difference. Calmness, curiosity, patience, and the right kind of questions can gradually change the entire atmosphere around mathematics at home.

The MathBridge Approach

The MathBridge approach is built on one simple idea: children learn mathematics best when emotional steadiness and mathematical understanding grow together.

A calm child who lacks understanding still needs instruction. However, a child who receives instruction while overwhelmed or panicked is unlikely to benefit from it. The brain's gates for learning are effectively slammed shut. The two must work hand-in-hand.

The goal of the approaches shared in this book is not aimed at perfection. It is aimed at steady, manageable progress:

- One calmer start to an evening session.
- One clearer explanation shared between parent and child.
- One constructively recovered mistake.
- One breakthrough moment where the child looks at a problem and says, “I can try again.”

How Language Shapes Children’s Mathematical Experience

The words adults use around mathematics do far more than describe a learning task. Over time, language patterns shape how children emotionally predict an experience before it begins. If mathematics is consistently associated with correction, urgency, unfinished work, frustration, or high-stakes performance, the brain begins preparing for stress before any actual thinking starts.

Children do not respond only to the worksheet directly in front of them. They also respond to what their nervous system expects is about to happen. Some children begin approaching their math tasks already anticipating correction, disappointment, tension, or failure. This anticipation quietly drains their attention, memory, emotional regulation, and basic willingness to engage before the learning itself has fully started.

Let’s look at the language “homework”. While the word "homework" may feel entirely neutral to an adult, for many anxious learners, it has slowly become emotionally connected to fatigue, unfinished tasks, domestic conflict, or the fear of getting things wrong. I encourage parent to shift from the phrase “homework” and move toward “home learning”. “Home learning” is not a magic phrase, and changing a single term does not instantly remove math anxiety. However, when paired with calmer interactions and realistic support, it helps reposition the experience around learning, growth, thinking, practice, and progress rather than simply completion or performance.

MathBridge encourages parents to gradually shift away from the phrase "homework" and move toward "home learning."

The Parent Language Bank

Language also shapes exactly where children place their attention. A phrase that sounds completely ordinary to an adult can carry heavy emotional weight for a struggling child.

Consider how common phrasing can inadvertently trigger a panic response, and how supportive language patterns reduce emotional threat to guide children back toward thinking.

THE PARENT LANGUAGE BANK

WHAT WE OFTEN SAY	WHAT AN ANXIOUS CHILD HEARS	THE SUPPORTIVE MATHBRIDGE ALTERNATIVE
"Complete your work."	A mountain of urgency, burden, or an invitation to fail.	"Let's work through just one part first."
"Hurry up."	Immediate pressure to prioritize speed over actual reasoning.	"Take the time you need to think carefully."
"You should know this already."	Absolute proof of personal, innate inadequacy.	"This idea still needs practice, and that is completely normal."
"Just focus."	A confusing criticism they don't know how to execute during cognitive overload.	"Let's reduce one thing at a time and find the first step."

The goal is not to make every single sentence perfect. The goal is to gradually create a calmer learning environment around mathematics in your home. Over time, these language patterns help children associate mathematics with thinking, effort, recovery, curiosity, and the internal confidence to remain engaged, recover from mistakes, and continue thinking when the learning becomes difficult.

Parent Insight
The aim is not to make mathematics easy all the time. The aim is to make difficulty feel manageable enough that your child can keep thinking.

BridgeKey™: MB-Q7X9

Chapter 2: Understanding Mathematical Development in Grades 3-5

Grades 3–5 represent a major shift in primary mathematics. In Grades 3-5, children move rapidly beyond early counting and simple number work into place value, mental strategies, regrouping, multi-digit multiplication, long division, fractions, measurement, time, money, estimation, word problems, multi-step reasoning, and formal mathematical explanations.

This stage is not simply a matter of children doing "more math." It is the exact point where children suddenly need to manage several different types of thinking at the same time. To solve an upper-elementary problem, a child must hold abstract information in their mind, select an appropriate strategy, organize their written steps, translate concepts into appropriate language or mathematical vocabulary, and critically check whether their final answer actually makes sense.

For many children, this feels as though the ground has suddenly shifted beneath them. The mathematics that once felt highly visible, intuitive, and manageable now demands memory, language, planning, and flexibility thinking, all within the span of a single task.

What This Shift Means for Parents

When a child begins to falter during these years, parents must realize that the struggle is rarely a matter of a lack of effort.

Consider how much thinking is required to solve a single word problem. To solve it successfully, a third, fourth, or fifth grader must read the text accurately, identify the relevant information, completely ignore unnecessary "distractor" details, plan a sequence of steps, choose the correct operation, calculate accurately, and then reflect on the logic of their result. This requires children to coordinate many different skills at the same time. Then, it becomes significantly harder if the child is simultaneously worrying about making a mistake.

This reality explains why a child may appear to "suddenly" struggle in the upper primary. The vulnerability isn't sudden. Gaps in foundational understanding often become visible only when the mathematics curriculum begins demanding this level of processing and managing the demands at once. A child who previously managed by simply memorizing basic steps now finds themselves frustrated and mentally exhausted. In such case, children do not need more pressure, they need support that addresses the real challenge.

THE GRADES 3-5 COGNITIVE SHIFT

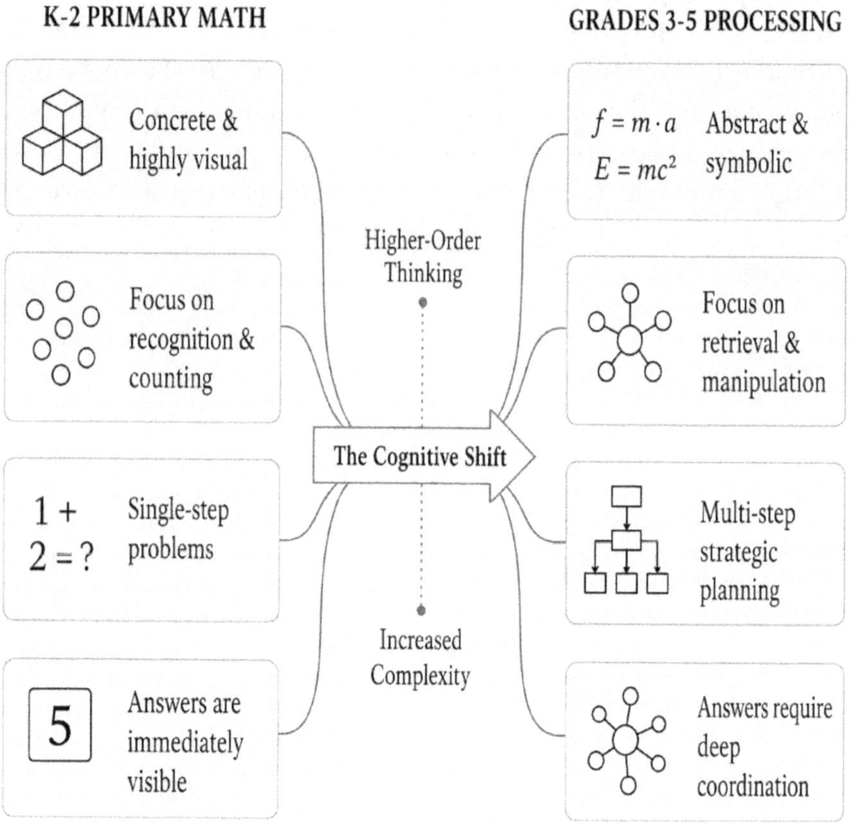

Parents must learn to recognize the signs of genuine cognitive exhaustion. Throughout my career, I have worked with children who began a math lesson appearing calm and capable, only to become frustrated and appear completely overwhelmed after solving just one or two problems. I openly acknowledge that, earlier in my career, I sometimes found myself thinking, "Come on, put in more effort." Over time, I realized that many of these children were actually protecting themselves from anticipated failure. Conversations with parents also revealed that they often perceived and interpreted this behavior as lack of effort when, in reality, the child had simply reached the point where they could no longer keep thinking through the task. For many parents, this realization can be uncomfortable. It is entirely possible to care about your child's experience and still feel frustrated when you do not understand what is happening.

You may notice your child starts a home learning session with high energy, only to become drastically slower, more irritable, or highly inaccurate after solving just a few problems. This downward spiral is rarely a sign of sudden carelessness or defiance. More often, it simply means the task has entirely drained the child's ability to think.

When this exhaustion hits, forcing a child to sit through a prolonged session is counterproductive. Shorter, highly focused intervals of practice often produce better learning outcomes than long, gruelling marathons that end in tears and frustration.

The Developmental Bridge

During these intermediate years, a child is actively moving from doing mathematics primarily with physical objects (concrete) toward thinking mathematically (abstract) without relying on physical objects. Concepts like fractions, adding or subtracting with regrouping, and long division require children to truly understand numerical relationships, not just memorize a sequence of steps. Nevertheless, this progression does not require children to abandon concrete supports before they are ready, as these tools help build the conceptual understanding that makes abstract thinking possible.

Some parents frequently worry that if their child continues to rely on physical manipulatives, drawings, number lines, arrays, or place-value blocks, it means they are weak in mathematics. This reflects a common misconception. I understand why parents worry about this. Watching your child continue to rely on "supports" that other children seem to have outgrown can feel unsettling. However, needing these supports does not necessarily mean a child is falling behind.

One of the conversations I have had with some parents involves answering the question, "Why is my child still using 'baby blocks' in Grade 4?" I have had to reassure them that engineers and mathematicians also use models. Concrete and visual models are not academic "crutches", they are scaffolding that aids the understanding of mathematical concepts and one of the ways children express their mathematical reasoning. They are what allow a child to connect abstract symbols to tangible quantities, patterns, and relationships.

In the adult world, mathematicians, engineers, architects, and data scientists routinely use models, sketches, and visual diagrams to organize their ideas and solve complex problems. Visual thinking is not childish. It is how many experts in various fields including mathematician solve difficult problems.

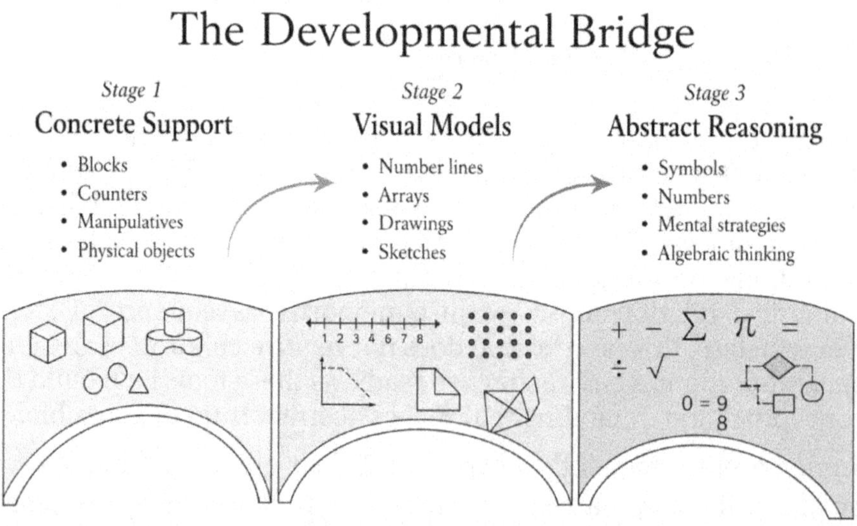

"Visual models are not academic crutches. They are the bridge between concrete experience and abstract understanding."

The critical balance for home learning lies in ensuring that these visual tools support independent thinking rather than replacing it. A child should not become permanently dependent on physical blocks for tasks they are developmentally ready to do mentally. The ultimate goal is a gradual, confident transition from concrete support to sketches and models, and finally to purely mental and symbolic reasoning (the concrete, pictorial, and abstract stages of learning mathematics). It is also important to remember that children may move back and forth between these stages as they develop understanding.

For parents, the practical takeaway is simple: avoid rushing your child away from making meaning.

If your child can successfully explain a complex fraction problem using blocks, sketches or a number line, that time is never wasted. That is the invaluable, quiet work of building mathematical structure. Once that structural meaning is secure, you can gently guide them across the bridge by asking: "Can you show me this exact same idea another way?" or "Can we try writing this out using only numbers now?"

Over time, children build sustainable math confidence not simply by getting answers right, but by developing an understanding that feels stable, connected, and meaningful. Practicing patience during this developmental transition protects far more than just a single evening's performance—it protects your child's fundamental willingness to keep thinking when the math gets challenging.

Parent Insight
When a child still needs a model, it does not automatically mean they are behind. It may mean they are building the bridge from visible understanding to internal mathematical thinking.

BridgeKey™: MB-Z3TN

Chapter 3: Understanding Math Anxiety

Math anxiety is far more than a simple dislike of mathematics. It is an intense emotional response that interferes with the thinking processes children rely on during learning.

Research on math anxiety consistently shows that stress affects working memory, attention, reasoning, and a child's ability to think clearly while solving problems. In practical terms, this means a child may temporarily become entirely unable to access thinking skills they are perfectly capable of using during calmer moments.

For many children, this anxiety begins long before they started solving math problems. The mere sight of a worksheet, the mention of a timed activity, an unexpected correction, a dense word problem, or hearing the phrase “It’s time for math” can trigger an immediate physical and emotional stress response. The brain begins preparing for failure before the mathematical reasoning process has had a chance to start.

Children are rarely consciously aware that this protective wall is going up. They simply experience math as something mentally exhausting, emotionally uncomfortable, and something they want to escape as quickly as possible.

The Myth of Sudden Forgetfulness

This invisible wall is exactly why math anxiety can feel so confusing for families. A child may understand a math strategy perfectly during class, or while working calmly with one-on-one support, only to suddenly appear forgetful, distracted, or highly resistant during home learning.

When this happens, parents often make logical but inaccurate assumptions. They assume the child wasn't listening in school, is not putting in effort, or never truly understood the concept in the first place.

The child's knowledge hasn't vanished. Instead, real-time anxiety is actively blocking their ability to organize, retrieve, and use what they already know.

Mathematics places a significant demand on working memory because it requires a child to hold multiple moving pieces of abstract information in their mind simultaneously. When anxiety occupies a significant portion of that mental space with worry, self-monitoring, and the fearful anticipation of mistakes, the brain is left with fewer resources available for thinking and problem-solving.

This explains why an anxious child's academic performance appears so frustratingly inconsistent. A child may solve a problem and explain a concept flawlessly on Tuesday evening yet struggle with an identical problem on Wednesday afternoon. The learning has not disappeared; rather, emotional tension has locked the gate to their retrieval system. This unpredictability is exhausting for both parents and children. However, recognizing it as an emotional block changes how we respond to it.

The Hidden Overload Pattern

Math anxiety does not always look like a dramatic meltdown. While some children do become visibly upset or tearful, others mask their panic through quiet, subtle shifts in behavior. I once worked with a student who tried very hard not to draw attention to herself during mathematics lessons. She remained quiet, calm and put on her "thinking face" in an attempt to mask her confusion and emotional overload. However, during individual sessions, I noticed that she constantly sought reassurance, apologized before attempting the simplest problems, and became visibly distressed after minor mistakes. Eventually, she explained that she spent much of every math lesson worrying that everyone would discover she was "actually bad at math."

Experiences like this are a reminder that math anxiety does not always look like avoidance, tears, or refusal. Sometimes math anxiety can be easy to miss, particularly when children work hard to conceal their distress.

Perhaps the most difficult part for adults is that children who appear to be coping well are sometimes the ones who need the most support.

When mathematics begins to feel emotionally threatening, a child's attention becomes divided. One part of the brain tries to process the task, while another part frantically monitors their emotional safety.

One way of understanding math anxiety is that it functions like a school fire alarm or fire drill. The classrooms, bathrooms, and admin office remain exactly where they have always been. The cafeteria has not disappeared. However, once the alarm goes off, the school's priority immediately shifts from learning to safety. In the same way, when a child experiences math anxiety, the brain temporarily shifts its priority away from problem-solving and towards emotional protection.

- Am I getting this wrong?
- What if I disappoint my parent?
- What if I look slow?
- What if everyone else understands this faster than I do?

This internal monitoring happens quietly and automatically, draining valuable thinking energy. On the surface, this cognitive overload manifests in several distinct behavioral patterns:

The Sprinted Guess (Rushing): For some children, rushing is a desperate defense mechanism. Slowing down increases their exposure to uncertainty and makes mistakes feel more visible. Finishing the page as quickly as possible feels much safer than sitting inside the discomfort of not knowing. Parents often misinterpret this speed as carelessness, when it is actually an attempt to mask emotional distress.

The Constant Reassurance Loop: Other children become entirely dependent on external validation, asking "Is this right?" or "Am I doing this correctly?" after almost every step. This rarely stems from a lack of understanding; rather, anxiety has lowered their trust in their own thinking. They require constant external confirmation just to feel safe enough to write the next step.

The Hidden Slate (Fear of Showing Work): Because a visible mistake can feel like an emotional exposure of inadequacy, an anxious child may erase their work repeatedly until the paper tears, refuse to write down intermediate steps, or insist on restarting a whole question from scratch after a minor error. They are reacting to the discomfort of seeing their mistakes visible on the page.

The Emotional Safety Threshold

Every young learner has an invisible threshold where emotional strain begins to interfere with their ability to think clearly. For children who have already experienced an ongoing frustration, classroom comparison, or the pressure of timed tests, this threshold is reached very quickly.

Once a child crosses this threshold, adding more academic explanation will not solve the problem. The difficulty is no longer mathematical, it is emotional.

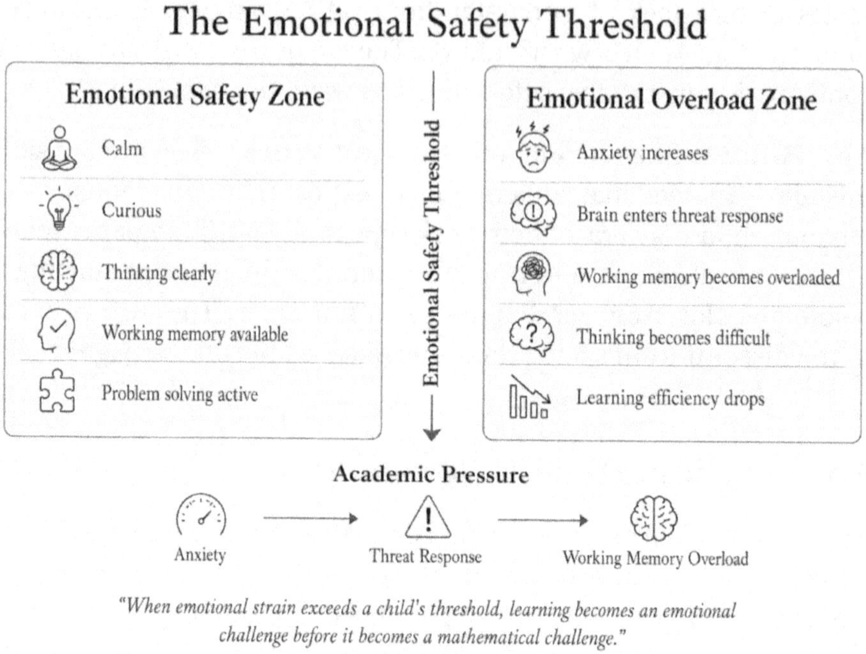

When a child crosses this line, they benefit from a deliberate reduction in complexity rather than an increase in parental urgency. They may need a brief pause, a physical movement break, a reassuring word, or a visual model to bring their nervous system back to a baseline where meaningful thinking can resume. This is not giving up or avoiding the work, it is a deliberate way of helping the child feel calm enough to think again and allows learning to continue.

Parents must also recognize that many children experience severe mental and emotional fatigue by the end of the school day. A child who is fiercely resistant during evening home learning is rarely being lazy. In most cases, they have simply spent their entire day devoting massive amounts of mental energy managing attention, social pressures, and learning hurdles in a classroom. They arrive at home learning already mentally exhausted.

The long-term goal is not to remove every academic challenge or make mathematics effortlessly easy. Children need to experience productive struggle to build resilience. The goal is to protect the emotional conditions that allow that thinking to happen.

Resilience is built not because math always feels comfortable, but because children gradually learn that confusion, errors, and challenges can be navigated without panic, shame, or emotional distress. The deepest support you can offer is not helping them find a correct answer faster—it is keeping them willing to stay at the table and keep thinking when the math gets tough.

Parent Insight
Before asking, "Why won't my child try?" it may be more useful to ask, "What feeling is my child trying to avoid?"

BridgeKey™: MB-L2KD

Chapter 4: Why Math Anxiety Becomes a Confidence Cycle

Math anxiety rarely develops through one isolated event. More often, it mutates into a repeating, self-sustaining cycle that slowly shapes how children think, participate, and feel about mathematics over time.

The cycle typically begins when a child experiences a wave of confusion, time pressure, consistent correction, classroom comparison, or repeated experiences of difficulty. The resulting emotional discomfort immediately makes their thinking less organized and more difficult. Because their thinking and problem-solving ability is strained by stress, the child naturally experiences more mistakes, slower processing, and growing hesitation around unfamiliar tasks. All of these experiences culminate in an increasingly negative association with learning mathematics.

Those negative experiences gradually begin reinforcing a destructive internal belief: “Maybe I am just bad at math.” Once this belief takes root, the next mathematical experience does not begin from an emotionally neutral starting point. The child now approaches home learning or classroom mathematics carrying heavy memories, negative expectations, and defensive anticipation. Before a pencil touches paper, they are already expecting confusion, frustration, embarrassment, or failure. This predictive anxiety effectively compromises their attention, working memory, and ability to think clearly.

Over time, the cycle becomes increasingly difficult to break. The child worries because mathematics feels difficult, and mathematics becomes more difficult because the child is worrying.

The Gradual Erosion of Participation

Notably, this cycle moves slowly. Confidence rarely vanishes after a single difficult worksheet or one poor test result. Instead, children gradually build their emotional expectations from cumulative daily experiences.

A child who often leaves a mathematics session feeling unsuccessful, heavily corrected, confused, rushed, or negatively compared to peers will eventually stop approaching mathematics as a place to actively think. Instead, mathematics becomes an emotional minefield—a place where the child must constantly monitor their own performance, frantically avoid mistakes, and protect their ego from feeling unsuccessful yet again.

The Confidence Cycle

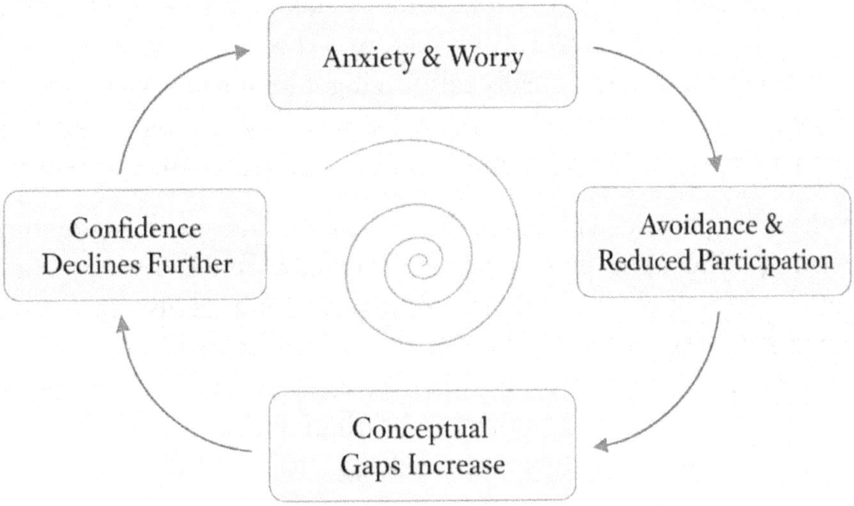

Repeated emotional experiences gradually shape how children think, feel, and participate in mathematics.

For parents, this gradual shift is critical because it fundamentally changes how fully a child participates in the learning process.

Children who expect absolute difficulty begin to deploy classic defensive maneuvers:

- They attempt fewer problems independently.
- They rely completely on adult prompting.
- They disengage and shut down at the first sign of a challenge.
- They become reluctant to try anything that "feels" unfamiliar
- They stop practicing the very skills that they need to strengthen.

Over time, this reduced participation creates gaps in academic understanding. Those gaps then make future grade-level mathematics feel harder than before. The cycle becomes both emotionally and academically reinforcing. Meaning, anxiety reduces participation, reduced participation weakens learning opportunities, and weaker understanding increases future anxiety.

Eventually, parents will notice this cycle extending far beyond home learning. The child may become unusually quiet during school mathematics lessons, hesitate to answer aloud, refuse to explain their reasoning to the teacher, or experience intense worry before assessments. Understanding this pattern allows parents to step in with strategic, early support rather than waiting until a child's confidence has completely bottomed out.

The Pressure-Avoidance Strategy

Once mathematics becomes associated with emotional discomfort, avoidance becomes a highly logical coping strategy for a child. Avoidance creates immediate, short-term emotional relief. If a child successfully delays starting a task, jokes around instead of trying, asks an adult for the answer immediately, or rushes to finish without thinking, the uncomfortable feeling of anxiety temporarily decreases.

The core problem is that this relief is short-lived. Over time, avoidance strips away the exact opportunities a child needs to strengthen their understanding, practice mathematical reasoning, recover from normal errors, and experience the psychological triumph of successful problem-solving.

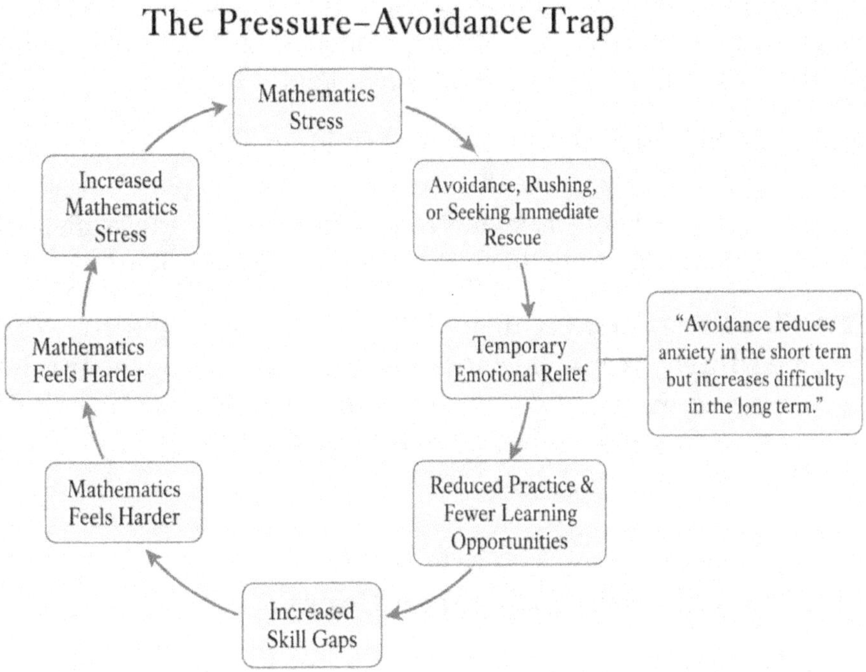

As mathematics becomes more complex in Grades 3–5, children also become fully aware of performance differences among their peers. They notice who answers questions instantly, who never seems confused, how little support their classmates require, and how frequently they themselves become stuck. These silent, daily comparisons quietly layer emotional tension over an already demanding subject.

At home, this cycle can become emotionally exhausting for parents. You may begin feeling confused, impatient, worried, or entirely unsure of how to help effectively. It is completely natural to fear that your child is losing motivation, becoming overly dependent, or "giving up too easily." Left unchecked, home learning interactions slowly transform from calm support into high-tension standoffs that leave both parent and child feeling discouraged.

To interrupt this, parents must look past the visible behavior and ask a completely different evaluative question: *What might this behavior be protecting my child from feeling right now?*

A child who refuses to begin a task is rarely just avoiding doing math, they are actively avoiding anticipated failure, public embarrassment, or severe cognitive overload.

The Perfectionist Mask: For some children, the cycle is driven by perfectionism. They erase their work repeatedly, panic over tiny errors, or restart whole problems after a minor slip because mistakes feel emotionally catastrophic rather than simply informational.

The Passive Mask: Other children disengage quietly. They become entirely passive during home learning, waiting for an adult to spoon-feed them every single step, or instantly weaponizing the phrase "I don't know" before giving their own thinking a chance to develop.

The Internal Story Children Carry

By ages 8–11, children are doing far more than just learning mathematics; they are actively constructing an internal identity narrative as learners. I recall, a few years ago, meeting one of my new students who had just moved into Grade 5. He introduced himself by saying, "Hi, my name is Fred, and I am not good at math." After spending several months working to build his confidence, I realized just how early children internalize their identities as learners. This internal narrative is shaped largely by the emotional experiences and learning environments they encounter over time.

Watch closely, and you can map the degradation of a child's internal story through the subtle evolution of their language:

"Can you help me?" to "I don't get this." to "I'm bad at math." to "I hate math."

For many children, these heavy statements are not casual complaints. They are deep-rooted, protective conclusions designed to lower the bar of expectation and reduce future disappointment, embarrassment, or emotional risk.

Consider a child who genuinely enjoyed numbers in the early primary years. As the curriculum shifts toward multi-step reasoning, more complex word problems, and abstract concepts, they encounter visible, daily struggle. They watch their peers breeze through tasks. During home learning, their frustration spikes the moment answers don't come quickly.

Slowly, without any single dramatic event occurring, the child's internal story quietly flips:

- *Old Story: "Math is something I am actively learning."*
- *New Story: "Math is a public tool that shows what I cannot do."*

Once this internal narrative solidifies, it can bleed into other areas of development. Children may become overly cautious in other academic subjects, less willing to take creative risks, quicker to abandon challenges, or increasingly dependent on adult reassurance across the board. When struggle is often tied to a child's identity rather than their growth, their willingness to learn begins to shrink.

Why Reassurance Alone Fails

When a child is in learning distress, a parent's immediate instinct—an instinct shared by many teachers and other adults, including myself early in my career—is to offer loving reassurance: "You are good at math!" or "Don't worry, you are so smart; you can do it!"

These words come from a place of deep care and a desire to shield a child's self-esteem. Reassurance is incredibly important for establishing emotional safety. However, verbal reassurance alone is unequipped to rebuild broken confidence when a child's daily, lived experiences communicate the exact opposite message.

If mathematics consistently feels confusing, overwhelming, and frustrating, the child will eventually trust their own experience over your verbal encouragement.

Internally, their brain asks a highly logical question: *"If I am so good at this, why does it feel like a nightmare every single time?"*

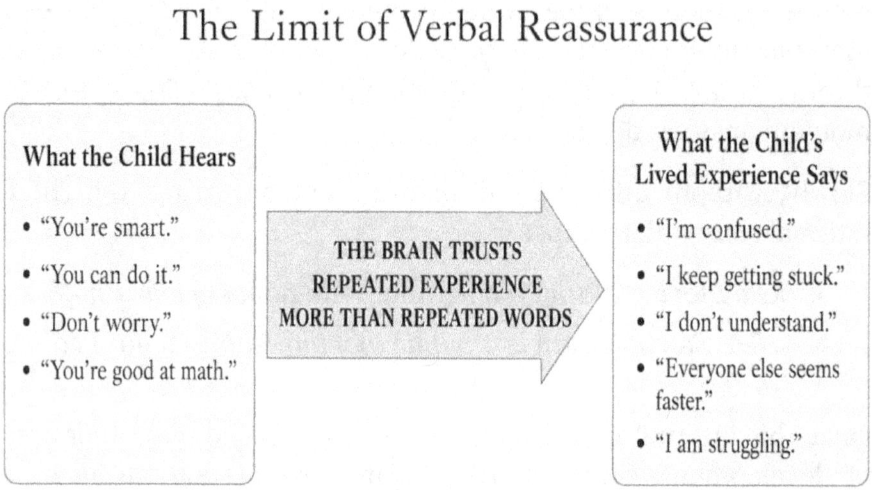

This does not mean encouragement is useless. Children absolutely thrive on calm, supportive language from the adults they trust. The core message here is that confidence is not built by what children are told, it is built by what they consistently experience while learning.

This principle extends well beyond mathematics. Children gradually come to trust what they repeatedly experience more than what they are repeatedly told. Encouragement matters. It helps children feel emotionally safe. However, encouragement alone cannot outweigh a pattern of repeated failure. Imagine a child learning to climb a wall. Each time they try, they slip and fall because they have not yet learned the right technique. At first, they may keep trying, especially if an encouraging adult cheers them on. However, over time, repeated falls begin to shape their expectations. Eventually, the child's own experience becomes more convincing than the encouragement. They start to believe, *"I can't do this"* and may stop trying altogether. The solution is not louder encouragement. The solution is helping the child experience small, achievable successes that gradually build both skill and confidence.

Rebuilding confidence requires pairing your encouragement with believable, undeniable experiences of progress, independent thinking, and successful recovery. Children begin trusting their own thinking again not because we shower them with compliments, but because they visually witness themselves navigating a challenge and surviving.

A quick success story to reinforce why praise and encouragement alone is not enough. I worked with a family where the parents shared with me that nothing seemed to be working with their child, no matter how much they encouraged him. There were plenty of words of motivation and reassurance.

When I first began working with the child, mathematics often triggered immediate distress. He frequently asked for answers before attempting to think independently and became frustrated and tearful whenever he got stuck. What struck me most was that he had already concluded that struggling meant he was “not good at math” and that there was nothing anyone could say to change his perception. His parents were doing everything they knew how to do: encouraging him, reassuring him, and trying to protect him from further frustration. Unfortunately, the more they reassured him, the less he seemed to believe them.

As we progressed through our sessions, we decided to focus first on building emotional safety, reducing performance pressure, and creating opportunities for small wins. We focused on making learning visible by documenting what he could not yet do and regularly comparing it with what he could now do independently. Eventually, he began asking almost every day, "Can we do math today?" He began to develop genuine excitement for learning mathematics. This experience reinforced my belief that encouragement, when paired with small wins and visible evidence of progress (making learning visible), has enormous potential to create a true turnaround.

Small Wins, Big Evidence

To provide your child with this undeniable evidence, you must look for and celebrate tiny, breakthroughs that have nothing to do with a perfect test or assessment scores:

The Isolated Step: For a child who usually shuts down when facing a complex word problem, a massive win looks like successfully identifying only what the question is asking, before attempting to solve anything else.

The Calm Correction: For a child who panics and tears the paper after an error, a win looks like calmly erasing a single digit and correcting it without rewriting the entire page.

The Independent Launch: For a child who refuses to work without you holding their hand, a win looks like independently reading and completing just the first step of a problem before asking for your guidance.

These moments may seem insignificant to an adult, but psychologically for children, they become the foundation on which confidence is rebuilt. They provide the mind with concrete, undeniable proof that math struggle can be managed, errors can be corrected, and thinking can continue when the work is challenging.

As a support system, your focus should shift beyond the final correct answer. A calmer start to the evening, a brief independent step, a faster recovery after frustration, or a willingness to explain a strategy aloud are the true signs that confidence is being rebuilt.

The deepest long-term goal is not to force a rapid spike in achievement through high-pressure drills. The goal is to fiercely protect your child's fundamental willingness to remain engaged with the learning process itself. Children who remain willing to try, question, and persist will naturally develop deep mathematical understanding over time. Protecting that willingness to keep thinking, even imperfectly, is the strongest foundation you can possibly build.

Parent Insight
Confidence is built through repeated experiences of success, not repeated messages of reassurance.

BridgeKey™: MB-R8PW

Chapter 5: What Parents Often See vs What May Actually Be Happening

During home learning, parents naturally respond first to what is highly visible. For example, they may see outright refusal, sudden tears, distractions, rushing, arguing, heavy silence, or complete emotional shutdown. These moments can feel incredibly frustrating, especially when you are already tired from your own day, deeply worried about your child's academic progress, or trying your best to provide support. Parents often tell me that these moments leave them feeling both heartbroken and exhausted.

The challenge is that behavior rarely communicates its true cause.

A child who appears careless is often a child who feels completely overwhelmed.

A child who looks completely unmotivated is usually a child who already expects failure before they begin the learning.

A child who argues constantly during mathematics is not simply being defiant, they are reacting to learning environment that they already associate with confusion, pressure, embarrassment, or intense emotional discomfort.

I encourage parents to pause and look beneath the surface behavior before reacting to what they see. This is not an invitation to excuse every poor reaction or strip away all academic expectations. Children absolutely require boundaries, responsibility, and consistent opportunities to persevere through challenges. However, we can only support that perseverance when we understand what the behavior is trying to communicate.

How to Read Behavior More Carefully

Children in Grades 3–5 rarely possess the emotional vocabulary or self-awareness to clearly articulate what is happening inside their minds during a stressful learning moment. A ten-year-old will almost never look up from a worksheet and say:

- “I am really afraid of getting this wrong and looking uncapable.”
- “I can't keep all these steps straight in my head.”
- “I already feel so far behind my peers that trying feels pointless.”
- “I am experiencing a wave of panic when I look at these mistakes.”

Instead, these feelings often show up as everyday behaviors that adults easily misunderstand.

The Behavioral Translation Matrix

To bridge this communication gap, parents should learn to translate the visible signs into their true underlying causes:

This matrix illustrates why home mathematics interactions derail so easily. As adults, we naturally focus on controlling what the child is doing externally, while the child is entirely consumed by surviving what they are feeling internally.

The Behavioral Translation Matrix

WHAT PARENTS OFTEN SEE OR HEAR	WHAT MAY ACTUALLY BE HAPPENING INTERNALLY
"I don't want to do this."	The child is anticipating confusion, failure, or intense emotional discomfort before the task even starts.
"This is boring."	The child is using boredom as a protective shield to avoid feeling unsuccessful, exposed, or inadequate.
"I forgot how to do everything."	Real-time anxiety has disrupted the child's retrieval pathways, locking away concepts they easily grasp when calm.
Rushing through problems.	The child is fleeing the situation. Slowing down forces them to sit within the discomfort of uncertainty, making mistakes feel more visible.
"Can you just help me?" (Instant ask)	The child has entirely lost trust in their own thinking and is seeking an emotional safety net before risking an independent try.
Tears, anger, or sudden shutdown.	The child's cognitive processor has crossed its threshold and is completely overloaded.
Clowning around or becoming silly.	The child is using humor to reduce the tension and escape a situation that feels threatening to their self-esteem.
Silence or staring blankly at the page.	The child is experiencing a cognitive freeze; they legitimately do not know where to begin and are paralyzed by the weight of the task.

The Ripple Effect of Our Interpretations

Your initial interpretation of a behavior is one of the most important factors to consider during home learning, because interpretation dictates how you respond.

If you interpret your child's avoidance strictly as laziness, your natural response will be to increase pressure, demand compliance, and introduce urgency. In return, the child feels less emotionally safe, their anxiety spikes, and their thinking and reasoning capacity shrinks further.

Similarly, if you interpret rushing purely as carelessness, your response will likely spiral into frustrated commands to "slow down." This completely misses the reality that the child is running a sprint to escape the emotional discomfort of uncertainty. If you view constant help-seeking only as annoying dependence, your response may turn into impatience or a sharp withdrawal of support—right when the child is crying out because they lack the basic confidence to think independently.

The Response Trap vs. The Support Route

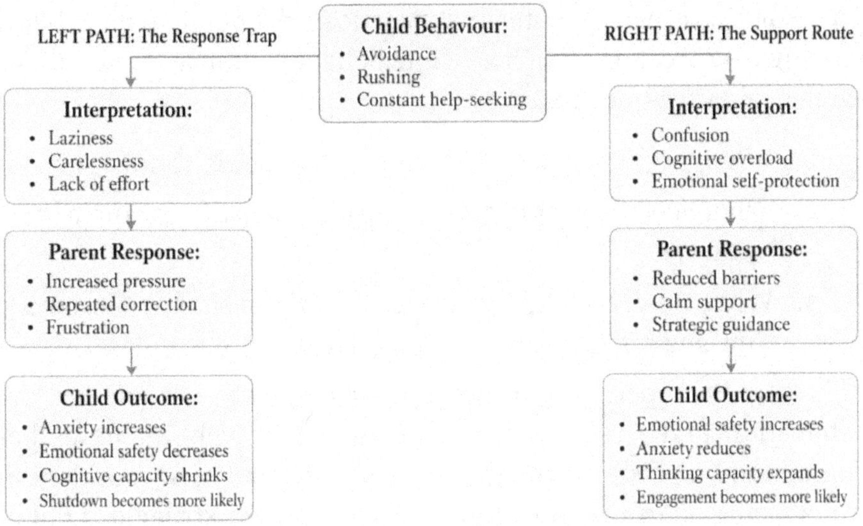

"The same behaviour can lead to very different outcomes depending on how it is interpreted"

When you consciously reframe these behaviors as red flags for confusion, overload, or emotional self-protection, your parental response naturally becomes calmer, more precise, and highly supportive of actual thinking. The mathematics still matters, but you are aware enough to remove the emotional barrier preventing your child from engaging with it first.

Left uninterrupted, accumulated misunderstandings create persistent, negative family patterns. Parents begin dreading home learning sessions, expecting bitter resistance before a book is opened. Children walk to the table already bracing for correction, frustration, and conflict. Both parties enter the interaction emotionally tense, rendering calm, logical thinking practically impossible for everyone involved.

The "Looking Beneath" Method

To break this cycle, parents must practice a deliberate pause. Before moving to correct or punish a visible behavior, stop and ask yourself three evaluative questions:

1. What might my child be trying to escape right now?
2. What specific part of this task feels too heavy for their brain to carry?
3. What small adjustment would help them safely re-enter a thinking state?

This cognitive pause prevents reacting strictly out of our own exhaustion and stress. When we focus exclusively on immediate compliance, speed, or finishing an assignment, you accidentally transform a learning opportunity into a power struggle. Looking beneath the behavior slows the interaction down enough to reveal where the breakdown is actually occurring.

Scenario A: The Word Problem Refusal

- **The Surface Behavior:** Your child instantly pushes a word problem away and says, "I'm not doing this." Your immediate reaction might be: "You aren't even trying to put in effort."
- **Looking Beneath:** You realize the child is overwhelmed by the dense block of text, cannot isolate the core question from the filler details, and already expects to fail.
- **The Pivot:** Instead of delivering a lecture on the value of hard work, you reduce the immediate cognitive weight. You say: "Let's ignore the math for a second. Let's only figure out what story this question is trying to tell us first." The academic expectation hasn't been removed; you have simply provided an accessible entry point.

Scenario B: The Mental Math Sprint

- The Surface Behavior: Your child flies through a sheet of mental math, scribbling down guesses and repeated errors. Your immediate reaction is to snap about lack of focus.
- Looking Beneath: You recognize that slowing down causes their internal worry to spike. They are terrified of appearing slow, making a mistake in front of you, or confirming their worst fears about their thinking ability.
- The Pivot: Instead of repeating the unproductive instruction to "slow down," you strip away the time pressure. You cover the page and say: "We are only going to look at this single problem right now. Take all the time you need. There is absolutely no rush." The standard remains high, but the emotional threat is relieved.

Scenario C: The Instant Help-Seeker

- **The Surface Behavior:** Your child reads a single sentence of a division problem and immediately whimpers, "Can you just do it for me?" You feel frustrated by their total lack of independence.
- **Looking Beneath:** The child is completely capable of doing the work, but their self-trust is so entirely fractured that they are outsourcing all of the cognitive risk to you. If the answer is wrong, it becomes your failure, not theirs.
- **The Pivot:** Immediately handing over the answer strengthens this dependency. Instead, you validate their safety while keeping them in the driver's seat: "Show me the very first tiny part you think you understand, and we will build the next step together from there."

Shifting the Family Climate

Children thrive when the adults in their lives approach their academic struggles with curiosity rather than judgment. This approach does not mean lowering the bar or removing accountability. Rather, it ensures that children experience difficult learning moments as stable situations that can be analyzed, supported, and systematically simplified—rather than moments that inevitably lead to shame, conflict, or personal defeat.

As adults, we intuitively understand this need for psychological safety. If you are a stressed employee who feels constantly monitored and criticized by a manager, you require reassurance, clear guidelines, and calm conditions before you can perform high-level analysis or solve a complex corporate problem. If you are a driver caught in an overwhelming, dangerous traffic jam, you become drastically more focused the moment the immediate noise and tension reduce.

Your child's brain operates under these exact same biological conditions. They will always think more clearly when they feel emotionally regulated enough to face a task, rather than emotionally cornered by it.

The Learning Atmosphere

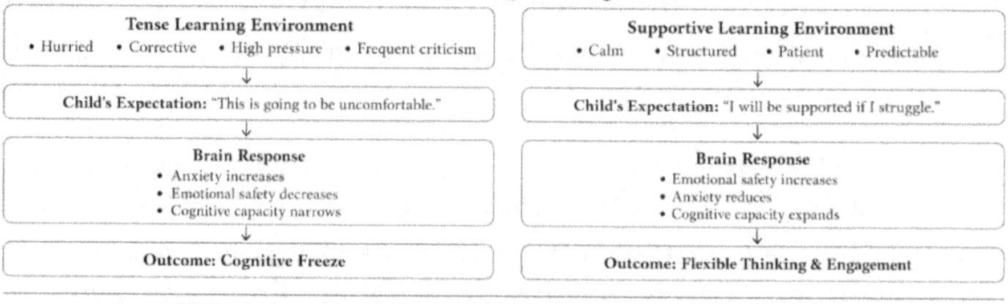

"Children think most effectively when they feel emotionally safe enough to engage with challenge."

Every single day, your child is forming an internal conclusion about what mathematics means emotionally within your family. If those interactions are consistently tense, rushed, and corrective, math becomes synonymous with domestic distress.

But when you consistently respond with a calm demeanor, clear structure, patience, and emotionally regulated support, your child learns a far more valuable life lesson: difficult challenges can be encountered, managed, and overcome without panic, humiliation, or emotional shutdown.

Moving forward, try to view your child's behavior not as an obstacle to correct, but as vital data about what they are struggling to process internally. One of the most profound lessons I learned during my postgraduate studies in Learning Support and Inclusion was that behavior often communicates something important. That idea truly changed the way I interpreted children's responses to learning. Rather than seeing behavior as something to fix, I gradually began to view it as a diagnostic tool. Doctors do not treat symptoms in isolation; they try to understand what is causing them. In the same way, a child's behavior is often not the real problem. Instead, it can provide valuable information about where they are experiencing difficulty, stress, or emotional overload.

This single perspective shift is the ultimate starting point for a sustainable, effective, and peaceful home learning environment.

Parent Insight
The visible behavior is not always the real problem. Sometimes it is the child's best available protection strategy.

BridgeKey™: MB-Y7LH

Chapter 6: How Confidence Breaks and How Parents Rebuild It

Math confidence rarely develops through a single successful lesson, one encouraging comment, or a good test result. More often, confidence grows gradually through consistent, everyday experiences that teach children something foundational about themselves as learners:

- "I can think."
- "I can recover."
- "I can stay with difficulty."
- "I can improve when something feels challenging."

In exactly the same way, confidence usually breaks down gradually rather than suddenly. Over time, children begin to expect mathematics to feel a certain way.

Small, fleeting moments that adults may barely notice can gather internally for a child: feeling rushed by a ticking clock, needing constant micro-correction, struggling without understanding why, feeling unable to keep up with peers, or leaving a home learning session emotionally exhausted far more often than successful.

Over time, these moments cause children to form expectations about what mathematics will feel like before a home learning session begins. At this point, mathematics is no longer the only challenge. The child is also trying to manage fear, self-protection, and the worry that today's lesson will confirm what they already believe about themselves.

Confidence matters because it shapes whether children stay engaged with learning. If there is one lesson that transformed my practice over the years, it is this: Early in my career, I believed that confidence followed achievement. Experience taught me that the relationship between confidence and achievement is far more complex. For some children, achievement does indeed build confidence. However, for many children who struggle with mathematics, the opposite is often true, achievement follows confidence. This can be difficult for adults to accept because our natural instinct is to focus on fixing the academic problem first.

Children who believe they can safely remain engaged during a challenge are far more likely to attempt unfamiliar problems, explain their reasoning aloud, ask clarifying questions, recover after mistakes, and tolerate the temporary confusion that is a natural part of learning.

In contrast, children whose confidence has been fractured begin prioritizing emotional safety over participation in learning. Confidence is not a soft extra or an optional personality trait; it directly determines whether a child remains willing to engage with the very learning experiences required for their understanding to grow. When a child pulls back from mathematical thinking to protect their feelings, they lose daily opportunities to build independence, flexible thinking, resilience, and recovery skills.

How Confidence Actually Breaks

Confidence weakens the moment a child stops experiencing mathematics as a place for active thinking and begins experiencing it strictly as a place of constant evaluation. Instead of focusing on curiosity, reasoning, tracking patterns, or problem-solving, the child's mind shifts entirely toward performance management, avoiding mistakes, protecting their self-worth, and trying strategies not to appear incapable.

Crucially, confidence does not break only through obvious criticism or outright academic failure. It can collapse through quiet experiences where a child's effort feels entirely disconnected from success.

A child may try incredibly hard over multiple weeks while still feeling repeatedly confused, dependent on an adult, slower than their classmates, or emotionally exposed during lessons. Over time, the child stops questioning just the math problem and begins questioning whether they can handle difficult challenges.

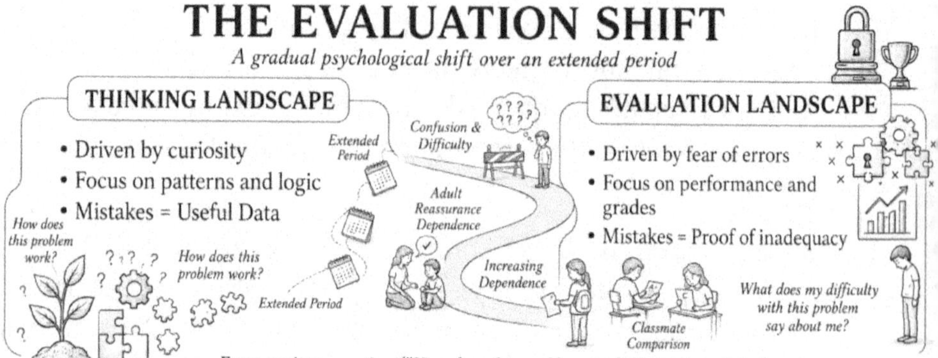

From curious question ("How does this problem work?") to internal doubt ("What does this difficulty say about me?"), a gradual and silent shift.

For many children, this shift happens in total silence. Some children become highly perfectionistic, spending far more effort trying to avoid mistakes than exploring mathematical ideas.

When children interpret academic challeneg as evidence of a personal deficit rather than a normal feature of the learning process, mathematical difficulty becomes tied directly to their identity instead of their growth. This toxic conclusion can eventually bleed into other areas of life. The child is no longer just protecting themselves from incorrect answers. As mentioned earlier, they are protecting their ego from the painful experience of feeling incapable.

The Rescue Trap

Sometimes, parents who begin by helping with "just one question" discover, 45 minutes later, that they have essentially completed Grade 4 mathematics themselves. When parents see their child discouraged, overwhelmed, or melting down, their natural instinct is to step in and resolve the frustration quickly. Offering immediate help feels caring, responsible, and highly efficient. In a high-stress moment at the end of a long day, completely taking over a task can feel like the kindest and most practical solution compared to watching a child continue to struggle.

However, rescuing a child too early sends a damaging, unspoken message directly to their subconscious: "You are not capable of managing this without me."

When a parent constantly steps in to clear the path, the child learns to sit back and wait for adult intervention before attempting any independent thinking. What began as a loving act of support slowly turns into learned helplessness. The child discovers that the discomfort of a difficult problem can be bypassed entirely by transferring 100% of the cognitive responsibility back to the adult.

I sometimes think about this in the same way we teach children to ride a bicycle. If we never loosen our grip, of course children remain safe, but they never discover that they are capable of balancing on their own. Confidence develops not because someone continues holding the bicycle, but because the child gradually experiences what it feels like to stay upright independently.

A parent explains one seemingly complex question to lower the stress level, then another and then the next one. Eventually, the child begins seeking help sooner and sooner, attempting less and less independently, and becoming increasingly distressed by the slightest hint of uncertainty, error, or temporary confusion around solving mathematical problems.

This leaves parents trapped in an exhausting emotional paradox:

"If I help too much, my child becomes entirely dependent on me."

"If I step back completely, my child becomes emotionally flooded and overwhelmed."

The way out of this trap is not to eliminate parental support, but to provide support without removing the child's ownership of the thinking process. Children build sustainable confidence by personally experiencing the loop of attempting, stumbling, correcting, and recovering during a challenge. If an adult consistently performs the heavy lifting, the child may finish the worksheet successfully, but they will quietly lose all trust in their own independent ability to function.

To break this, embrace the principle of Gradual Release of Responsibility. A parent can reduce immediate complexity, offer one smaller step, ask a targeted guiding question, model solution to a problem, or help organize the visuals on the page before gently returning the cognitive weight back to the child. This ensures the child experiences the safety of your support while retaining total ownership of learning.

The Productive Struggle Window

To learn, children absolutely must encounter a certain amount of mental struggle. However, it is vital to recognize that not all struggle is built equal.

- Productive Struggle occurs when a child experiences a genuine challenge while maintaining full cognitive access to their thinking. The child may feel stretched, uncertain, or mentally pushed, yet they remain regulated enough to reason through options, test out strategies, ask targeted questions, and stay actively engaged.
- Emotional Overwhelm looks entirely different. Once emotional overload spikes past a certain threshold, the brain's reasoning systems become disorganized and inaccessible. The child drops into a survival state—panicking, shutting down, guessing randomly, refusing to continue, or looking for any possible avenue to escape the task entirely.

One of the most important roles you play as a parent is acting as a thermostat to keep your child inside that optimal Productive Struggle Window: challenged and supported, but never emotionally flooded.

Learning breaks down at both extremes of this spectrum. Tasks that are too easy require zero meaningful processing or persistence, while tasks that are overwhelmingly difficult trigger a wave of anxiety strong enough to lock down working memory.

Parents often unintentionally push children out of this optimal window in both directions. Over-helping strips away the opportunity for productive thinking too quickly, while applying excessive pressure, urgency, or too much abstract complexity all at once drops the child straight into emotional overload.

The Productive Struggle Window

Learning grows when children experience enough challenge to think, but not so much that they become overwhelmed.

Too Easy	Productive Struggle	Overwhelmed
• Little thinking required • Minimal challenge • Limited learning • Confidence changes very little	• Thinking is active • Mistakes are expected • Persistence develops • Confidence grows through success	• Working memory becomes overloaded • Anxiety increases • Thinking becomes disorganized • Children often give up or seek rescue
If the task is too easy Increase the level of challenge slightly.	**If the child is in productive struggle** Stay nearby. Encourage thinking. Avoid taking over.	**If the child becomes overwhelmed** Reduce the complexity. Break the task into smaller steps. Restore emotional safety before continuing.

The goal is not to eliminate struggle. The goal is to keep children inside the Productive Struggle Window where thinking, confidence, and learning grow together.

You can actively regulate this balance by adjusting the size and pacing of the task rather than removing the challenge completely. This might mean:

- Slowing the pace of the learning session.
- Folding the worksheet to show only one visible question at a time.

- Deploying a single, targeted visual support (like a number line or array).
- Reviewing a highly familiar, successful example right before tackling a new problem.
- Asking a guiding question to help the child move past the immediate sticking point.
- Allowing a scheduled pause aimed at emotional regulation.

If a child is drowning in a fractions assignment, they do not need the entire worksheet thrown in the trash. Instead, they benefit from working through a single, carefully selected example with you, slowly enough to experience success and understanding, before attempting the very next problem on their own.

This process teaches a child an invaluable emotional lesson: difficulty is something that can be survived.

True resilience is built when children watch themselves stay engaged through a manageable challenge and emerge successful on the other side, rather than escaping the discomfort immediately or being broken by it. Over time, this builds a high tolerance for uncertainty, errors, and temporary confusion. It prevents them from automatically interpreting a difficult moment as proof of personal failure.

The Confidence Before Speed Principle

Many well-meaning adults without realizing it, communicate a damaging myth, that being successful at mathematics means being fast. Children are highly perceptive. They notice instantly who finishes their schoolwork first, who shouts out answers immediately, who rarely appears stuck, and how adults lean in with visible impatience when an answer takes longer than expected.

Over time, children internalize this and equate speed with innate capability. This reshapes how a child approaches their learning. Instead of focusing on deep reasoning, pattern recognition, conceptual flexibility, and true understanding, they pivot entirely to performance management, frantically asking themselves:

- “How fast can I finish this page?”
- “How quickly can I avoid making an error?”
- “What do I need to do to avoid looking slow?”

When speed becomes the primary metric of success, mathematics stops being an exploration of thoughtful problem-solving and becomes an exercise in stress management. This spikes anxiety around errors, completely suffocates curiosity, discourages creative risk-taking, and makes children intensely defensive about trying unfamiliar strategies.

As a parent or educator, you must remember a core truth: rapid performance and deep conceptual understanding are not the same thing.

Many children process mathematics carefully and slowly because they are actively organizing the information in their minds, cross-checking numerical relationships, and forming a rock-solid conceptual foundation. Conversely, a child may appear fast simply because they are relying on rigid rote memorization or guessing to escape the emotional discomfort of the task as quickly as possible.

While automatic recall of mathematical facts and processing efficiency are important for higher-level mathematics, fluency built entirely through fear, urgency, or intense adult pressure is incredibly fragile under stress. Children who tie their self-worth to speed struggle immensely the moment the curriculum advances to complex topics that demand flexible, creative reasoning rather than rapid recall alone.

Your focus should shift from how quickly your child answers to how securely they understand. A child who thinks methodically, clearly explains their reasoning, catches and corrects their own errors, and remains emotionally engaged during a challenge is building an infinitely stronger long-term mathematical foundation. Conversely, a child who sprints through their work while remaining anxious and detached are in the danger of developing foundational of conceptual understanding.

Real, sustainable confidence grows when a child learns that in your home, thoughtful thinking is valued more than a rushed performance.

Re-Defining Success Together

One of the most powerful ways to rebuild math confidence is to rethink what success looks like during home learning.

When confidence is low, many children begin measuring success using only one standard: getting the right answer. If the answer is wrong, they feel unsuccessful. If they need help, they feel unsuccessful. If they make a mistake, they feel unsuccessful.

Over time, this creates for children, a situation where mathematics becomes a constant test of self-worth rather than an opportunity to learn.

Parents often unintentionally reinforce this pattern by focusing heavily on completion, accuracy, speed, or grades. While these outcomes certainly matter, they are not the only indicators of meaningful progress.

Children build confidence more effectively when they experience success through actions they can control rather than outcomes they cannot always guarantee.

One powerful way to rebuild confidence is to involve your child in helping define what success will look like before a home learning session begins.

For example, rather than asking, "Can you get all of these correct?" you might ask:

"What would a successful learning session look like for you today?"

At first, many children will still focus entirely on answers:

- "Getting everything right."
- "Finishing quickly."
- "Not making any mistakes."

These responses provide an opportunity for a valuable conversation. Together, you can gradually broaden their definition of success.

Success might look like:

- Starting the work without arguing.
- Attempting a challenging problem before asking for help.
- Explaining their thinking aloud.
- Staying calm after making a mistake.
- Trying a second strategy when the first one does not work.
- Completing part of a task independently.
- Identifying where they became confused.
- Remaining engaged for a set amount of time.

Notice that none of these goals lower expectations. The mathematics remains exactly the same. The challenge remains exactly the same. What changes is where the child places their attention.

Instead of constantly asking, "Am I getting everything right?" they begin asking, "Am I doing the things that help me learn?"

This shift is important because confidence grows from consistent experiences of competence and ownership. When children help define achievable markers of success, they experience more frequent evidence that they are capable learners.

As their confidence grows, the definition of success can gradually evolve. A child who once celebrated staying calm during a difficult problem may later celebrate solving a multi-step task independently. The expectations increase naturally because the foundation beneath them has become stronger.

Remember, the goal is not to lower standards.

The goal is to create enough success experiences that your child remains willing to engage with the learning process.

Children who believe they can make progress are far more likely to persist through difficulty than children who believe success is reserved only for perfect performance.

In many cases, the most important question is not:

“Did my child get everything right today?”

It is:

“Did my child leave this learning experience believing a little more in their ability to learn?”

That belief is often the bridge that carries them toward future success.

> **Parent Insight**
> *When children help define success, they begin collecting evidence that they are capable learners*

BridgeKey™: MB-T4YM

Chapter 7: Understanding Your Child's Math Confidence Patterns

From Internal Experience to External Patterns

Earlier chapters explored what happens internally when mathematics begins to feel threatening for children. Anxiety divides a child's attention and makes it more challenging for them to think. It can make an otherwise manageable math problem feel much harder than it really is. Parents cannot see working memory becoming overloaded or confidence beginning to weaken. What they usually see instead is behavior.

Some children sharpen their pencil several times before beginning their work. Once they finally decide the pencil is sharp enough, the verbal responses begin. You may hear comments such as, "I don't understand" before they have read the question. Other children rush through their work, repeatedly erase their answers, or suddenly become frustrated.

These behaviors are outward clues to an experience taking place beneath the surface.

However, the connection between a child's internal experience and outward behavior is rarely straightforward. There is no single way children respond when mathematics feels emotionally draining. Two children can face exactly the same learning challenge and react in completely different ways. One may cry and refuse to continue. Another may become unusually quiet. A third may begin joking, changing the subject, or distracting everyone around them.

What makes these behaviors harder to interpret is that the same child may respond differently from one day to the next. A child who works confidently on Monday may need constant reassurance on Wednesday. These changing responses often leave parents wondering how best to help.

For this reason, confidence cannot be understood from one behavior alone. Hesitation does not automatically mean anxiety. Rushing is not always carelessness. Asking for help does not necessarily mean a child lacks independence. Refusal does not always mean defiance. To understand what is really happening, you need to look beyond the behavior itself. Consider the task, the learning environment, your child's previous experiences, and what usually happens next.

The question is therefore not simply:

What is my child doing?

A more helpful question is:

What might this behavior be helping my child manage right now?

A child who rushes may be trying to escape an uncomfortable experience. A child who repeatedly asks for help may be protecting themselves from the possibility of being wrong. A child who jokes, changes the subject, or leaves the table may be creating distance from a task that threatens their sense of competence.

To be clear, not every difficult behavior has a hidden psychological explanation. Children sometimes rush because they are tired. They become distracted because the room is noisy. They may resist because they would rather be doing something else. The goal is not to search for hidden meanings in every action. The goal is to stay curious enough to tell the difference between an isolated moment and a recurring response to mathematical challenge.

That distinction leads to one of the most important ideas in this chapter.

A behavior tells you what happened today. A pattern tells you what has been happening over time.

Recognizing patterns also protects parents from drawing conclusions too quickly. Children can have strong mathematical ability and still struggle to show it when anxiety, cognitive overload, or low or lack of confidence interfere with learning. Some think critically and solve challenging problems with ease, yet freeze under pressure. Others understand different calculation strategies when they feel calm but lose confidence in their own thinking as soon as anxiety takes over. This is why you should never judge your child's math confidence or ability by speed, compliance, or test scores alone.

This chapter is not about labeling your child or putting them into neat boxes. Instead, it is a guide to help you notice the behavioral patterns that may be shaping their relationship with mathematics.

Children are constantly changing. Their responses change as their learning experiences change. The same child may move between different behavioral patterns depending on several interacting factors, including:

- The immediate learning environment
- Their current emotional state and energy levels
- The difficulty and presentation of the learning task
- Previous experiences with a particular mathematical topic
- The amount and type of support available from adults

Looking for Patterns, Not Labels

Parents naturally respond to the behavior they see. When a child delays getting started with a math task, the parent encourages them to begin. Over time, the behavior can become the label, "You're so slow." When a child rushes, the parent reminds them to slow down. The label becomes, "You're not thinking carefully." When a child repeatedly asks for help, the parent encourages more independence. Before long, the child may simply be seen as "too dependent." Each response may seem reasonable when viewed on its own.

The difficulty is that isolated behaviors rarely tell the whole story.

Imagine your child refuses to begin mathematics on Monday, rushes through several problems on Wednesday, and asks for reassurance after every step on Thursday. When you view these separately, those behaviors may look like defiance, carelessness, and dependence.

When viewed together, they may reveal one underlying pattern: your child is trying to reduce the emotional risk of thinking independently.

This is why patterns matter. They connect behaviors that might otherwise seem unrelated.

Reading a pattern involves more than noticing something your child does repeatedly. It means paying attention to three simple questions.

1. What usually happens before the behavior?

Does it happen when the worksheet looks overwhelming? When the task feels unfamiliar? When your child is working against the clock? Or when you begin correcting mistakes?

2. What does your child do?

Do they withdraw, rush, seek reassurance, argue, repeatedly erase their work, guess, become unusually quiet, ask to stop, or begin to cry?

3. What usually happens afterward?

Does the task suddenly become easier? Do you take over? Does the difficult question get skipped? Does your child receive immediate reassurance? Does the learning session end?

Looking at these three questions will change the way parents interpret behavior. Your child may not deliberately choose avoidance or dependence. However, if a particular response repeatedly brings relief, reduces uncertainty, or transfers responsibility to an adult, it can gradually become their preferred way of protecting themselves.

Patterns also reveal important exceptions. A child who appears dependent during written calculations may reason confidently with concrete materials, drawings, or mental strategies. A child who shuts down during homework may participate well when there is no time pressure. A child who rushes through a worksheet may stay fully engaged while solving an open-ended math puzzle.

These exceptions matter because they remind us that the difficulty does not always lie entirely within the child. Sometimes the pattern is connected to a particular math topic, the way the task is presented, the learning environment, or the type of adult support available. Knowing where a behavior does not occur can be just as informative as knowing where it does.

Treat patterns as information, not conclusions. They help you build a clearer picture of how your child is currently experiencing mathematics.

Over time, repeated protective responses can affect much more than a child's willingness to complete mathematics. Children may gradually begin to question their own thinking. They stop trusting answers they once knew, abandon useful strategies after a small mistake, or feel unable to take a single step without seeking reassurance from an adult.

At that point, the difficulty has moved beyond a particular worksheet or mathematical topic. It has begun to affect a child's trust in their own thinking.

Parent Reflection Checklist

As you work through this checklist, resist the temptation to judge your child based on one difficult evening. Instead, look for behaviors that appear again and again over several weeks. The goal is not to label your child or measure ability. It is to understand the patterns that may be shaping your child's relationship with mathematics.

PARENT REFLECTION CHECKLIST

This checklist is designed to help you notice **recurring emotional-learning patterns** over time, not to diagnose ability. Observe across several weeks, not just one difficult session.

BEHAVIOUR TO OBSERVE	NEVER	SOMETIMES	OFTEN
1 Avoids mathematics whenever possible.	○	○	○
2 Rushes through work without looking for underlying relationships.	○	○	○
3 Says things like, "I'm bad at math."	○	○	○
4 Becomes highly emotional (tears, anger, or intense avoidance) during mathematics.	○	○	○
5 Asks for answers immediately without first attempting a solution.	○	○	○
6 Gives up or shuts down after making a mistake.	○	○	○
7 Freezes when faced with multi-step or complex problems.	○	○	○
8 Frequently compares themselves with others based on speed or scores.	○	○	○
9 Prefers only easy or highly familiar mathematical tasks.	○	○	○
10 Becomes unusually upset, defensive, or discouraged when corrected.	○	○	○
11 Hides confusion through silence, joking, or distraction.	○	○	○
12 Needs repeated reassurance before attempting the first step.	○	○	○
13 Avoids explaining their thinking or strategy aloud.	○	○	○
14 Becomes mentally exhausted much more quickly during mathematics than other subjects.	○	○	○
15 Panics when unsure which strategy or method to use.	○	○	○
16 Restarts an entire problem after making one small mistake.	○	○	○
17 Waits for an adult to direct every step instead of thinking independently.	○	○	○

*Parents may wish to make a copy of this checklist and use it across the weeks to **look for a pattern.***

No single tick-box tells the whole story. What matters is whether several of these behaviors appear repeatedly over time. As those patterns become more consistent, they often point to something deeper than a difficult worksheet or a bad day. They may reflect changes in how your child has begun to think about themselves as a mathematics learner.

When Children Stop Trusting Their Mathematical Thinking

Sometimes a child reaches the correct answer and still looks straight at you before writing it down.

"Is this right?"

You may respond, "What do you think?"

The child shrugs.

"I don't know."

Yet they knew enough to find the answer.

The uncertainty lies somewhere else. They do not fully trust their own judgement.

Trusting their own mathematical thinking matters because mathematics asks children to make decisions constantly. They decide where to begin, which information matters, what strategy might work, whether an answer makes sense, and what to change when an idea leads nowhere. A child should not depend on an adult to make every one of these decisions. At some point, they must learn to treat their own thinking as a reliable source of direction.

Self-trust makes that possible. Parents strengthen self-trust by protecting opportunities for children to make mathematical decisions instead of making those decisions for them.

Making mathematical decisions does not mean children always get the answer right. It means they use what they know, check their thinking, respond to new information, and correct mistakes when they discover something is wrong.

A child with self-trust might say:

"I think multiplication will work here because the groups have the same size."

"I'm not sure yet, but I'll try drawing it."

"That answer looks too large. I need to check something."

"I made a mistake in this step, but the rest of my strategy still makes sense."

Notice what these statements have in common. The child remains the thinker. The goal for parents is to support the thinking without taking over the role of the thinker.

A Consultation Observation

During one of my parent consultation sessions, I observed a parent helping their child solve a subtraction problem. Part of the session involved watching how they naturally worked together before discussing ways to strengthen the child's mathematical confidence.

The child was solving 68 – 23. They independently chose a decomposition strategy and correctly subtracted 20, arriving at 48. Then they paused.

For a few seconds, they simply looked at the page.

Before the child had a chance to continue, the parent said,

"Just subtract 3 from 48"

The suggestion was mathematically correct, and it came from a genuine desire to help.

The issue was not the explanation itself. It was the timing.

The child had already selected an appropriate strategy and successfully completed the first step. The remaining challenge was deciding what to do next. That brief pause was not necessarily a sign that the child was stuck. It was the visible part of thinking.

Parents sometimes interpret hesitation as confusion. Sometimes it is. Just as often, however, hesitation is the moment children organize their thinking, check whether their strategy still makes sense, or decide what to do next.

Those few seconds matter.

Each time children work through that uncertainty themselves, they gather evidence that they can rely on their own mathematical thinking. Each time an adult consistently provides the next step before children have the opportunity to decide for themselves, they may be receiving a different message:

"When I become uncertain, someone else does the thinking"

Over time, children begin expecting the adult to make the next decision for them. Instead of asking themselves, "what makes mathematical sense?" they begin looking outward for reassurance.

Parents strengthen self-trust not by removing every moment of uncertainty, but by learning to recognize the difference between productive thinking and genuine confusion. Sometimes the most helpful response is not another explanation, it is giving children a little more time to think.

Building Mathematical Self-Trust

Self-trust grows when children remain responsible for making, testing, and revising mathematical decisions.

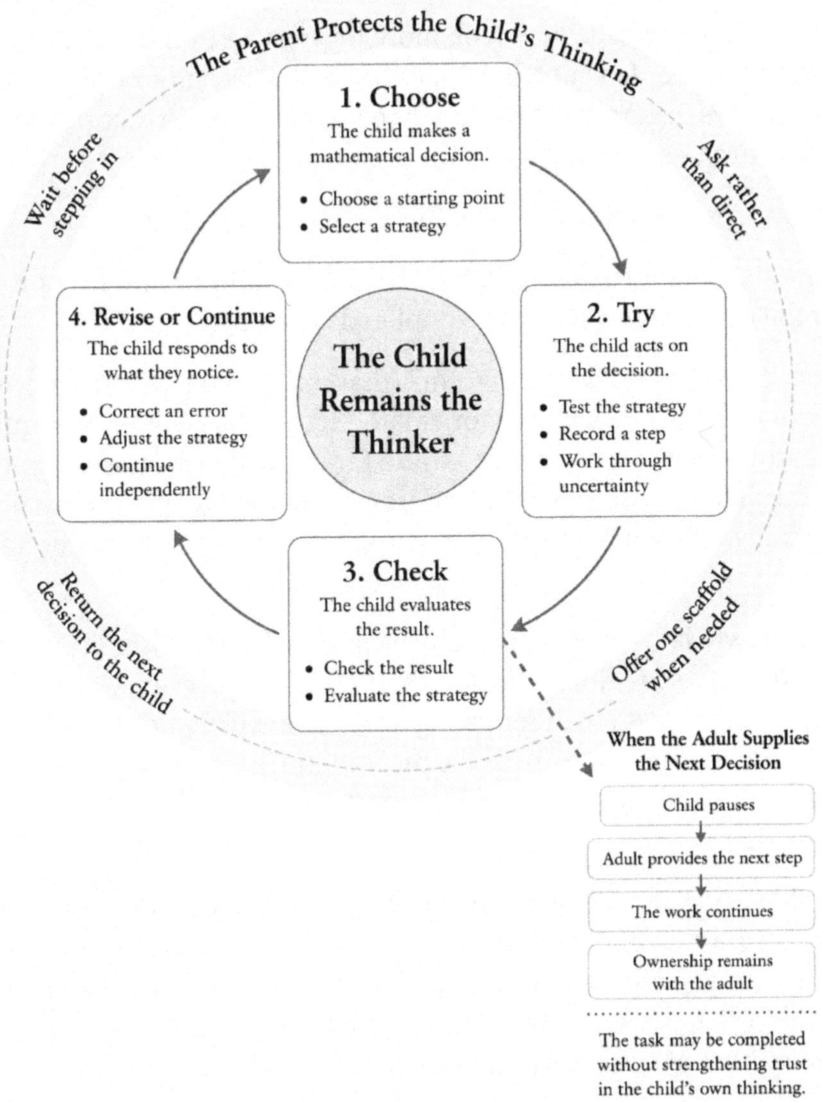

This distinction helps explain why completing more work does not always build more confidence. A child can complete ten questions correctly with constant direction from parents. Yet, leave the table with less trust in their own thinking than when they began. Another child may complete only three questions but make meaningful decisions, recognize an error, adjust a strategy, and explain their reasoning.

The completed page may look more impressive in the first example. But the deeper learning is happening in the second example, where the child is beginning to trust their own thinking and take ownership of the process.

Meaningful learning happened in the second instance.

Parents should therefore value the quality of the thinking as much as the number of questions completed correctly.

The consultation reminded me that rebuilding self-trust rarely requires adults to explain more mathematics. More often, it requires creating space for children to think for themselves. Those moments of hesitation, uncertainty, and revision are not interruptions to learning. They are often where the most important learning takes place.

I acknowledge, parents naturally want to help. They want to prevent frustration, save time, and see their children succeed. Yet effective support does not mean removing every obstacle. It means protecting the child's opportunity to make meaningful mathematical decisions, when those decisions involve uncertainty, mistakes, or changing direction.

Children develop self-trust by making mathematical decision. Every opportunity to choose a strategy, evaluate an answer, recognize an error, or explain their reasoning provides evidence that their own thinking has value. Over time, those experiences become the foundation of sustainable mathematical confidence.

As you continue reading, remember one simple question whenever you sit beside your child during mathematics:

Who is doing most of the thinking?

The answer to that question often tells you whether you are building your child's confidence and self-trust or unintentionally taking over the thinking.

Parent Insight
Patterns matter more than isolated moments. One difficult home learning session does not define your child. Consistent patterns provide the clearest insight into the support your child needs.

BridgeKey™: MB-H6QV

Chapter 8: When Parents Carry Their Own Math Anxiety

Many parents unknowingly bring their own experiences of learning mathematics into the home learning environment. Some remember the pressure of timed tests, public correction at the chalkboard, painful embarrassment, confusion, or the feeling of being behind. Others experience a surge of anxiety simply because today's teaching methods look entirely unfamiliar compared to the often, rigid ways they learned mathematics themselves.

Those experiences do not disappear simply because we become adults. Instead, they stay beneath the surface, quietly shaping an adult's tone, patience, body language, unspoken expectations, and the sudden urge to avoid, or correct a child's work too quickly.

A parent may never say a single negative word openly about mathematics to their child, yet the child can still instantly read the tension in the room. They sense it through a subtly accelerated pace, a tightened facial expression or a sharp shift in support tone,

Beneath these micro-interactions, parents are often carrying several hidden worries:

"What if I explain this concept incorrectly and ruin their learning?"

"What if my child falls behind their peers?"

"What if I don't understand the math well enough to help my child?

"What if this current academic struggle becomes permanent?"

When a parent interaction with a child is driven by these worries, children often perceived those worries and over time has an impact on the learning environment

Carrying math anxiety does not make you a harmful, incapable, or unsupportive parent. It makes you human. The critical question we must ask is whether that anxiety has quietly taken the driver's seat during your interactions with your child.

Recognizing the Worried Parent Profile

When anxiety begins driving interaction, a parent's supportive intentions can gradually turn into protective habits that unintentionally limit a child's thinking. The goal shifts from helping children develop confidence to helping them avoid difficulty.

Parents may explain before children have had time to think. They may step in the moment a child hesitates, not realizing that the pause often reflects active thinking rather than confusion. Some take over the first step before the child has had tried. Others correct every small mistake as soon as it appears, leaving little chance for children to notice and correct their own errors. Some parents move in the opposite direction, avoiding mathematics altogether because previous experiences have made every homework session feel emotionally exhausting.

Although these responses differ, they often have the same purpose: reducing immediate discomfort. In doing so, they can also reduce opportunities for children to develop resilience, independence, and confidence in their own mathematical thinking.

Research on home learning consistently shows that the emotional state parents bring to learning matters far more than explaining every mathematical step perfectly. Children benefit most when adults remain emotionally regulated, ask thoughtful questions, support the child's autonomy, allow time for thinking, and respond to mistakes without creating additional pressure.

Children learn as much from your emotional responses as they do from your mathematical explanations. They notice whether mistakes create tension, whether uncertainty is treated as a normal part of learning, whether struggle feels manageable, and whether the learning environment remains emotionally safe enough for them to continue thinking. For this reason, one of the most powerful ways to support your child's mathematical confidence is to become aware of your own emotional responses to mathematics. The calmer and more confident you become in the face of uncertainty and mistakes, the more likely your child is to see mathematics as a subject they can explore, persist with, and ultimately trust themselves to learn.

Modeling a Different Response

Recognizing your own math anxiety is an important first step. The next is learning to respond as you support your child's learning. You do not need perfect mathematical knowledge or complete mastery of the current curriculum. Your role is to model calm, flexible thinking when neither you nor your child immediately knows what to do. Children often adopt some of their emotional responses to challenge from the adults supporting them.

Small changes in language can gradually change the emotional atmosphere around mathematics:

Instead of saying, "I was never good at math either," say, "Math felt difficult for me sometimes too, but we can work through this strategy together."

Instead of saying, "That's easy," say, "This step feels tricky right now. Let's slow down and look at it carefully."

Instead of taking over the thinking, ask, "Which part of the problem makes the most sense to you so far?"

Instead of saying "That answer is not correct" with frustration. Say "Is this answer reasonable enough to accept"?

Before stepping in, pause and ask yourself this question:

Am I calm enough to support my child's thinking, or am I reacting from my own stress?

To conclude, children do not need parents who explain mathematics perfectly every time. They need parents who make learning feel safe enough for thinking to continue. Every calm response to a mistake, every moment of patience during uncertainty, and positive body language helps shape the way children experience mathematics.

Over time, those moments become the foundation of confidence, resilience, and trust in their own thinking.

Years from now, your child is unlikely to remember today's worksheet or every strategy you worked through together. They are far more likely to remember how mathematics felt while learning beside you. By bringing calm instead of anxiety, curiosity instead of urgency, and encouragement instead of fear, you help your child build a relationship with mathematics that extends far beyond the classroom.

> **Parent Insight**
> *You do not need a perfect relationship with mathematics to support your child. You need a healthier one than the one you inherited.*

BridgeKey™: MB-K9XN

Chapter 9: Making Support Decisions in the Moment

By this point in the book, you have explored how confidence develops, why emotions shape mathematical thinking, and how everyday interactions influence the way children see themselves as learners. Understanding these ideas is only half the challenge. The other half is applying them as you support your child's home learning.

Research across mathematics education, child psychology, behavioral science, executive functioning, and special education points to an important conclusion. Effective home support depends less on **how much** parents help and more on **how** they help.

Parents who encourage independence when support is needed are more likely to foster stronger mathematical engagement than those who frequently direct, correct, or take over. This distinction matters because increased involvement alone does not necessarily improve learning. When support becomes overly controlling, it can unintentionally reduce the very confidence parents are trying to build.

Children who experience repeated difficulty often need more support, and this naturally encourages parents to become increasingly involved. Yet greater control can also make children more dependent on adult guidance, creating a cycle in which each reinforces the other. Breaking that cycle requires more than good intentions. It requires thoughtful decisions about when to step in, what kind of support to provide, and when to step back again.

As discussed in Chapter 8, parents' own experiences with mathematics influence these decisions. Adults who feel anxious about mathematics may unintentionally become over-controlling, offer answers too quickly, or remain involved longer than necessary. These responses usually come from a desire to help, yet they can reduce opportunities for children to develop confidence in their own reasoning.

Effective support is also rarely about using more strategies or more resources. Adding multiple visual models, manipulatives, or explanations at the same time does not automatically increase understanding. Each support should address a specific challenge, requiring parents to become more intentional in the decisions they make. The goal is not to surround children with tools but to remove the barrier preventing them from moving forward.

Most importantly, effective support helps children become increasingly aware of their own thinking. It helps them identify what they already understand, recognize where their reasoning changes, select appropriate strategies, and reflect on whether those strategies are working. As children become more skilled in these processes, they become less dependent on adult guidance. The goal is not for children to explain every thought they have. It is to help them gradually direct and take greater ownership of their own learning with increasing confidence and independence.

This chapter builds on the ideas developed throughout the book by focusing on one practical question:

How do parents make effective support decisions during home learning?

The sections that follow present a framework for observing children's thinking, identifying the kind of support a situation requires, providing that support, recognizing when to withdraw it and helping children reflect on what they have learned. Together, these decisions create the conditions in which math confidence can continue to grow.

Read the Child Before You Read the Mathematics

Over the years, one of the most common mistakes I've seen parents make during home learning is assuming they already know why their child is struggling. A child who stops working is often described as distracted, careless, or confused. Yet the same behavior can arise for entirely different reasons. This is not to suggest that parents do not know their own child. They do. Parents understand their child's personality, learning history, strengths, and many of the challenges they have experienced over time.

However, math confidence is rarely lost because of one single experience. It gradually develops through many individual learning situations, each leaving an impact on how children see themselves as mathematician. If a series of learning experiences contributed to a child's loss of confidence, rebuilding that confidence must also happen one learning situation at a time. Rather than responding to what we assume is the child's overall problem, effective support begins by understanding the specific barrier the child is facing in this moment. Each response to support becomes one more positive learning experience, gradually rebuilding confidence over time.

Effective support begins by gathering information before offering it.

Rather than asking, "do you understand'? " or beginning a new explanation, start by examining what your child has already done. The work on the page often reveals more than the answer itself. It shows where thinking was progressing, where it changed direction, and where it stopped. Before deciding how to help, take a moment to understand what the work is telling you. Those observations will guide the support your child needs at that moment.

What mathematical decision comes next?

Once you have observed your child's work, resist the temptation to explain the entire problem as a default support. Instead, ask yourself a question:

What mathematical decision is my child trying to make right now?

Every mathematical task consists of a sequence of decisions rather than one large leap from question to answer. Before solving a problem comes first, decision making. Children must decide how to interpret the situation, which information matters, which strategy is likely to work, and whether their answer makes sense. When a child stops working, it is often because one of these decisions has become unclear.

Parents naturally focus on the final answer because it is the most visible outcome. Effective support, however, focuses on the decision that comes immediately before it. When we identify the specific decision that has interrupted a child's thinking, our support becomes more precise and often much less intrusive.

For example, a child solving a word problem may not need help with the calculation itself. They may be trying to decide whether to add or subtract. Another child may already know which operation to use but, because the problem is multi-step, be unsure where to begin. A different child may have started with an appropriate strategy but become stuck halfway through because they are no longer sure what to do next. Another may simply be trying to work out what the question is actually asking. Although these children all appear to be "stuck," each requires a different type of support because each is facing a different mathematical decision.

This small shift changes the parent's role. Rather than asking, "how can I explain this problem?" the more useful question becomes, which mathematical decision is preventing my child from moving forward?" Once that decision becomes clear, parents are far less likely to over-explain or provide unnecessary help. Instead, they can offer support that addresses the immediate obstacle while allowing the child to remain responsible for the thinking that follows.

Matching Support to the Barrier

Identifying the barrier is only the first decision. The second is deciding how to respond.

Two children may be struggling for exactly the same reason but may need different forms of support. Likewise, two children may receive exactly the same support though they are struggling for different reasons. Effective support depends not only on identifying the barrier but also on choosing the response that is most likely to remove it while preserving independent thinking.

One of the most common misconceptions about helping children learn mathematics is the belief that more help automatically leads to more learning. It doesn't. Effective support is not measured by how much adults explain or how long they sit beside a child. It is measured by whether the support removes the immediate barrier while allowing a child to continue thinking independently.

Sometimes, the best response is to do nothing for a little longer. This may seem counterintuitive. As a teacher, I find it one of the most difficult pedagogical decisions because everything in us wants to help the moment a child struggles. However, one of the realities of effective teaching is knowing when not to intervene. If your child is still thinking, testing ideas, or trying different strategies, resist the urge to interrupt. Silence often feels uncomfortable, but productive thinking is rarely instant. Stepping in too quickly can interrupt reasoning that simply needed more time to develop.

In other situations, the barrier has little to do with the mathematics itself. For example, your child may know how to solve the calculation but overlook the words *"how many more"* and add instead of subtracting. They may stop halfway through a multi-step problem because they answered the first question without realizing there was a second one. In these situations, reading the problem together, discussing unfamiliar vocabulary, or asking your child to explain the question in their own words may be all that is needed. Clarifying the task before explaining the mathematics helps ensure that your support addresses the real barrier rather than one you have assumed exists.

Some barriers arise because the thinking demands of the task have become increased. For example, multi-step problems require children to hold information, make decisions, and monitor their progress at the same time. Breaking the problem into smaller parts or recording intermediate steps can reduce unnecessary cognitive load while leaving the mathematical thinking firmly in your child's hands.

There are also times when a different representation provides the support your child needs. A number line, place-value chart, array, fraction model, or simple sketch can make an important mathematical relationship more visible. The goal is not to introduce as many tools as possible but to choose the one that best addresses the barrier your child is experiencing.

Occasionally, the barrier is not one of decision-making but of knowledge. Your child may not yet understand the mathematical concept or procedure that they need to solve the problem. In these situations, explicit teaching is appropriate. A brief explanation, worked example, or guided practice can provide the knowledge needed before your child resumes independent practice.

There are also moments when continuing to work on math for the day or working on a particular problem is no longer the best decision. Throughout this book, we have explored how cognitive overload affects learning. When working memory becomes overloaded, children have fewer mental resources available to process information, make decisions, and monitor their thinking. At that point, more explanations and more practice often become less effective, not more. A short break, a change of activity, or returning to the task later can allow cognitive resources to recover so that learning can continue more successfully.

The goal is not to use every type of support. It is to choose the response that best matches the barrier in front of your child. When support is carefully matched to the challenge, children receive the help they need while continuing to develop the confidence and independence to think mathematically.

If these ideas feel unfamiliar, that is completely normal. Like any new skill, learning to identify barriers and match the right type of support takes practice. You will not always get it right the first time, but with each learning experience, you will become more confident in making more effective support decisions.

Know When to Stop Helping

Support is most effective when it helps children regain ownership of their thinking. Once the barrier has been removed, the parent's role begins to change. Continuing to guide every step may feel helpful, but it can unintentionally replace the child's thinking with the parent's.

One of the most difficult decisions parents make is knowing when enough support has been provided. It is natural to want to stay involved until the entire task is finished. However, the goal of support is not to complete the work together. It is to help your child reach the point where they can continue independently.

As soon as your child begins making decisions again, consider stepping back. They may still work slowly, make mistakes, or need time to think, but these are often signs that they have resumed ownership of the learning. Remaining available for encouragement or occasional questions is different from continuing to direct every step.

Sometimes parents continue helping because they worry their child will become frustrated again. Others stay involved because finishing the task quickly feels more efficient. While these responses are understandable, too much support can gradually reduce opportunities for children to develop independence, perseverance, and confidence in their own mathematical thinking.

This does not mean withdrawing support completely. It means gradually reducing it as your child's independence increases. This gradual release of support aligns closely with the principle of student agency. A growing body of educational research shows that when children are given meaningful opportunities to make decisions about their learning, they develop greater independence, stronger self-efficacy, higher motivation, and a deeper sense of ownership over their progress. Crucially, these principles apply not only to typically developing learners but also to children with learning support needs. While the level of support may differ, every child benefits from opportunities to think, make decisions, and contribute actively to their own learning.

Practically, this does not happen all at once. Children rarely move from needing significant support to complete independence in a single lesson or over a few weeks. As their understanding and confidence grow, the amount of support should gradually decrease while the child's responsibility for the thinking gradually increases.

Educational research describes this process as scaffolding. Support is provided when it is needed and gradually removed as the child becomes more capable of completing the task independently. The scaffold is never the goal. It exists to help children reach a point where they no longer need it. Every time parents step back at the right moment, they create another opportunity for their child to think, make decisions, and strengthen the independence that supports them in learning mathematics.

A useful question to ask yourself is, "Can my child take the next step without me?" If the answer is “objectively” yes, allow them to do so. They may not solve the remainder of the task perfectly, but every independent decision strengthens their sense of agency, helping them become more confident and self-directed learners.

Effective support begins with understanding what is preventing a child from moving forward.

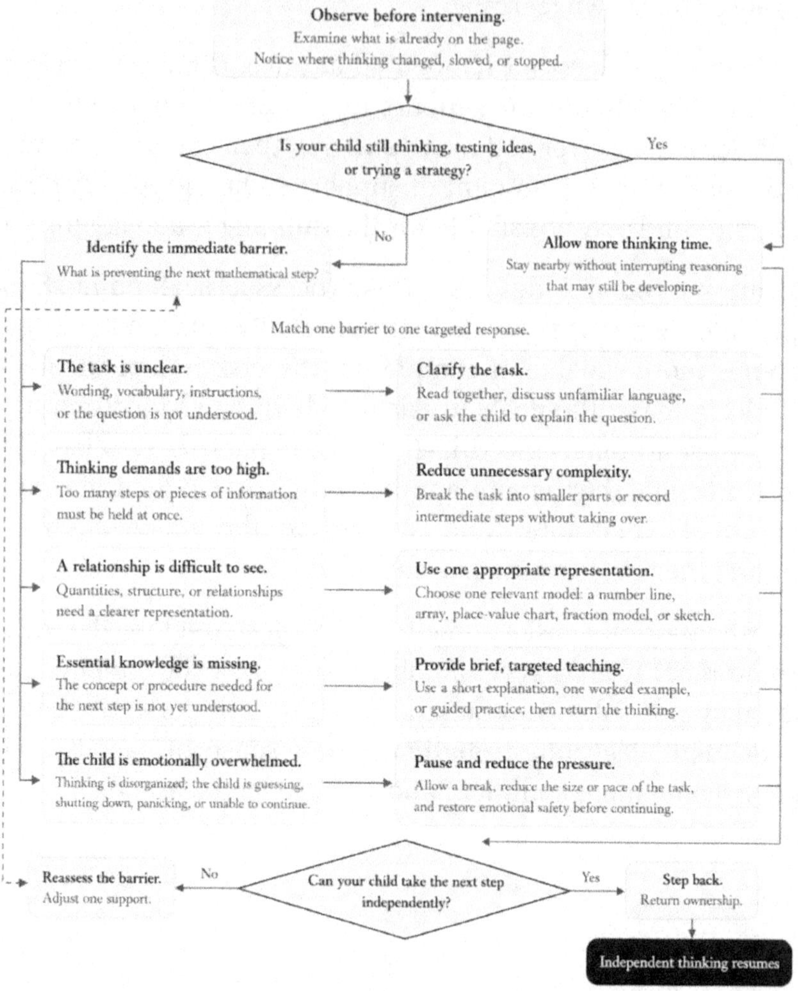

Observe first · identify the barrier · match the support · step back when your child is ready

BridgeKey™: MB-V3RD

Epilogue: A New Story for Math Success

As you finish this book and begin putting these ideas into practice, I hope you take a moment to notice not only what your child has learned, but also how they have changed as a learner.

Across more than two decades in education, working alongside children and families, one pattern has remained remarkably consistent form their success stories. Meaningful mathematical growth rarely begins with a better worksheet, a faster method, or a clever explanation. It begins when the conditions for learning change.

What stands out to me about these success stories is that they rarely happened because someone discovered a shortcut. They happened because adults approached learning differently. They remained consistent. They created emotionally safer learning experiences. They gave children repeated opportunities to think, recover, and try again.

This process is often slower than families hope, but it is far more powerful than they expect. Some children will make visible progress quickly while others may need more time to dismantle their anxieties and rebuild confidence. Every learner arrives at the table with a different history, different strengths, and unique challenges. What matters most is not comparing your child to an abstract milestone or another student. What matters is helping your child move forward from exactly where they are standing right now.

That is why I encourage parents to look beyond what can be measured immediately. Whenever you find yourself worrying about your child's math ability or waiting for higher scores as proof of progress, pause for a moment. Remember that some of the most important growth is invisible. Each time you choose calmness over pressure or offer one manageable hint instead of taking over the pencil, you are helping your child build self-trust.

Before you begin the Parent Support Journey, I encourage you to hold onto these three foundational ideas:

1. Give the process time.
2. Choose calmness over pressure.
3. Trust your child's capacity to grow.

These ideas may appear simple, but together they shape the kind of learning environment in which confidence grows. You do not need to create flawless learning experiences, nor do you need to know every current mathematical strategy. Your job isn't to remove every barrier. What your child needs most is an environment where they feel able to keep thinking, mistakes remain survivable, and support remains emotionally safe.

If your child leaves this journey believing a little more in their own ability to think, recover from mistakes, and persevere through difficulty, then a profound transformation has already begun.

Long after they forget specific worksheets, strategy, or practice pages, they will remember something far more powerful. They will remember how it felt to navigate a challenge in your presence, and they will remember believing that they could keep going.

If there is one idea I hope stays with you, it is this:

Protect your child's willingness to keep thinking. Everything else grows from there.

When children learn that they can keep going when learning becomes difficult, they carry that lesson far beyond mathematics. It shapes how they approach challenges, uncertainty, and mistakes throughout life.

The Parent Support Journey

This section brings the ideas from the Foundational Guide into practice. Each week follows a familiar rhythm, so you know what to look for, how to support your child, and how to build confidence, independence, and mathematical understanding over time.

Over the years, I have found that anxious learners thrive when the adults around them are calm, predictable, and supportive. The same is true for parents. They often feel more confident when they have a clear and reliable support structure. The words we choose during moments of confusion, frustration, mistakes, or uncertainty can shape how children experience mathematics.

Before beginning Week 1, take some time to review **The Parent Language Guide**. You do not need to memorize the examples or use them perfectly. Instead, use the Parent Language Guide as an ongoing reference throughout the journey whenever difficult learning moments arise.

After the **Parent Language Guide**, you will begin nine weeks of carefully selected learning experiences.

Weekly Purpose

The key emotional and mathematical goals for the week.

Parental Learning Focus

Practical guidance to help you better understand your child's experience and strengthen supportive learning habits at home.

The Science Behind This

Research-informed explanations that help explain why the strategies and approaches support confidence, learning, and long-term retention.

Companion Portal (Resources)

Printable activities, games, templates, and additional resources designed to reinforce the week's learning focus.

Using the MathBridge Companion Portal

Throughout this book, you will find QR codes that provide access to the MathBridge Companion Portal. To unlock the learning activities, you will need the **BridgeKey™**, which is built into this book. You will find BridgeKey™ code prompts at different points throughout the book. Simply follow the instructions on the portal and enter the requested BridgeKey™ to gain access to the learning activities.

The portal is organized to match the Parent Support Journey. Each week has its own section, and every learning activity follows a simple numbering system. For example, the resources for Week 1 are labeled 1.1, 1.2, 1.3, and so on. This makes it easy to locate the activities that connect with each week's learning.

I recommend reading each weekly chapter before opening the companion portal. Take time to understand the purpose of the week, reflect on the ideas, and consider how they relate to your child. When you reach the end of the chapter, scan the QR code to access that week's learning activities. The activities help you apply the ideas from the chapter while supporting your child's mathematical understanding and confidence.

Although the journey is organized into weeks, you do not need to follow a strict schedule. Some families may complete a week's learning in a few days, while others may spend more time exploring the same ideas. Move at a pace that works for your child and your family. Building mathematical confidence takes time, and meaningful progress grows through small, consistent steps.

Return to the MathBridge Companion Portal whenever you want additional practice or wish to revisit a concept. The goal is not to complete every activity as quickly as possible. The goal is to understand the ideas, reflect on your child's learning, and use the companion activities to support meaningful mathematical growth.

The book provides the understanding. The MathBridge Companion Portal provides the opportunities to put that understanding into practice. Together, they help you build your child's confidence, strengthen mathematical understanding, and create positive learning experiences at home.

Week 1: Understanding Your Child & Building Emotional Safety First

Weekly Purpose

This week is about changing how your child experiences mathematics the moment they sit down to begin. Before worrying about mistakes, accuracy, or gaps in understanding, the goal is to help your child feel safe enough to stay with the work.

Success this week isn't measured by speed or getting every answer right. It's measured by a calmer child who is willing to stay with the work when it feels difficult.

MathBridge Reminder

"Observation should come before correction."
"A child's behavior during math often tells a story."
"Children think more clearly when they feel safe enough to make mistakes."

◈ Parent Learning Focus

When a child experiences math anxiety, a blank math page no longer looks like a page of numbers. It feels like a threat. Before they read the first question, their stress response can take over. When that happens, thinking clearly becomes much harder.

In my work with families, the biggest mistake I see parents make during week one is trying to fix the math errors. If your child writes 12 instead of 21, or completely forgets a basic fact they knew the previous day, do not correct them yet. Interrupting them to fix an error while their anxiety is high instantly spikes their stress and makes it much harder for them to use the thinking skills they already have. For Week 1, prioritize the emotional routine over perfect answers. Once they feel safe sitting at the table, they will have more room to think clearly and begin noticing those mistakes on their own.

Your focus this week is to become a scientist observing your child's anxiety triggers. Watch their physical cues, notice when their shoulders tighten, when they hold their breath, or when they start tapping their pencil frantically. We are shifting from being a "home learning corrector" to an "emotional anchor." We'll use highly accessible, low-stress math activities to practice a new way of approaching mathematics before it becomes more challenging.

□ Science Behind This

When a child experiences math anxiety, the brain can begin treating numbers as a threat instead of a learning opportunity. The amygdala becomes highly active and temporarily disrupts the prefrontal cortex–the part of the brain responsible for reasoning, problem solving, and working memory. As a result, children may freeze, forget strategies they already know, or rush through problems without thinking clearly.

As children experience manageable success, stress decreases and confidence begins to grow. This helps them think more clearly, engage more willingly, and persevere through increasingly challenging mathematics.

Continue Your Learning Online

This week's learning activities help you build your child's mathematical confidence while creating a calm and emotionally safe learning environment. As you work together, encourage curiosity, celebrate small successes, and strengthen place value understanding through supportive learning experiences.

Scan the QR code to open the BridgeKey™ Companion Portal.

BridgeKey™: MB-P7LT

Week 2: Deepening Emotional Safety & Place Value Mastery

Weekly Purpose

This week is about helping your child understand how our number system works. A strong understanding of place value makes mathematics much easier. `Children begin to see how numbers are organized instead of trying to memorize them. The goal is to move beyond simply reading numbers and help your child understand what each digit actually represents.

MathBridge Reminder

"Place value is not a rule to memorize. It is a structure to understand".
"Each digit has value because of where it is, not just what it is".
"Understanding place value makes larger numbers easier to reason about".

◈ Parent Learning Focus

Many children memorize procedures without truly understanding quantity relationships. In Grades 3 through 5, your child is moving rapidly from comfortable 3-digit numbers into larger ones, 4-, 5-, and 6-8digit numbers. A child might read a large number like 352, 418 correctly aloud, while still struggling to understand the true value of the individual digits. For instance, they might look at that in the ten-thousands place and think it only means "fifty" or "five," missing that it actually represents fifty thousand.

As numbers become larger, children can no longer rely on simply counting what they see. They must understand what each digit represents. This shift toward more abstract thinking becomes increasingly demanding because the visual cues disappear and the strain on working memory increases.

When children make mistakes with big numbers, parents sometimes interpret those errors as simple carelessness or rushing. More often than not, your child's brain is simply overwhelmed by the visual noise of the digits. Your focus this week is to help your child slow down and see the "hidden identity" of numbers before they ever attempt to add or subtract them. We want to treat big numbers like a blueprint to unpack, rather than a giant wall of symbols to fear.

□ Science Behind This

Think of your child's thinking space as a small desk. When too much information arrives at once, it quickly becomes crowded. A child looking at the number 352,418 without understanding place value often sees six separate digits competing for space. Instead of seeing one organized number, they have to work much harder to keep track of each individual digit.

This can overwhelm the child before the actual calculation begins. Teaching place value helps children group digits into meaningful parts, such as thousands and ones. Instead of seeing six separate pieces of information, they begin seeing one organized number. That leaves far more room for calculation, reasoning, and problem solving.

Continue Your Learning Online

This week's activities build on the foundation you established in Week 1. Continue developing your child's confidence while deepening place value understanding and encouraging flexible thinking with larger numbers.

Scan the QR code to open the BridgeKey™ Companion Portal.

BridgeKey™: MB-F8WC

Week 3: Building Confidence Through Games & Strategic Thinking - Mental Math

Weekly Purpose

We are helping children move from counting-based strategies to more efficient and flexible ways of thinking about numbers, such as building to the nearest ten or compensating. The goal is to give your child enough time to think. Games provide repeated opportunities to practice these strategies until they become more natural.

> **MathBridge Reminder**
>
> *"Games reduce pressure because mistakes become part of play."*
> *"Confidence grows when children recover from mistakes."*
> *"Strategy matters more than winning."*

◈ Parent Learning Focus

In the upper-elementary grades, children who rely entirely on written column methods for simple calculations quickly hit a wall. When a child has to stop everything to manually write out or finger-count a basic step like 14 - 9 while working on a long-division problem, they often lose track of the rest of the problem.

Your focus this week is to recognize that mental math is not about speed, it is about strategy. Games help lower the pressure because children become more willing to experiment with different ways of thinking instead of worrying about getting every answer right.

Many children panic because they believe good mathematicians calculate instantly. When they don't know an answer right away, they guess just to end the awkward silence. This week, we use games to turn that silence into an opportunity to think. We want to show them how to break numbers apart (for example, seeing 98 + 45 as 100 + 43) and give them the emotional space to think through those changes without an adult rushing them.

□ Science Behind This

Working memory is the part of the brain that temporarily holds information while we think. It has limited capacity. When children have to spend too much of it working out basic calculations, they often lose track of the bigger problem they are trying to solve. Instead of thinking about the strategy, much of their attention is spent on simple arithmetic.

Mental math strategies help reduce that load. As children learn to group numbers into manageable parts, they no longer have to think about every single digit separately. This frees up working memory, making it easier to reason, solve problems, and notice mistakes.

Continue Your Learning Online

This week's learning activities encourage your child to think strategically rather than rely on memorization alone. Explore mathematical games and mental strategies together while creating opportunities for discussion, reflection, and independent thinking.

Scan the QR code to open the BridgeKey™ Companion Portal.

BridgeKey™: MB-N5ZH

Week 4: Multi-Digit Addition & Subtraction Toolkit

Weekly Purpose

This week builds on everything your child has learned so far. We now begin applying those ideas to larger addition and subtraction problems. Rather than relying on traditional column procedures that often require carrying or regrouping without a clear understanding of why they work, we introduce a flexible toolkit of mental strategies, including partial sums, compensation, constant difference.

MathBridge Reminder

"Understanding comes before procedures".
"Flexible thinkers become stronger problem-solvers".
"Understanding reduces fear".

◈ Parent Learning Focus

In school, most of us learned a single, strict approach to multi-digit addition and subtraction: stack the numbers vertically, line up the place-value columns, and work from right to left. If a column totals more than nine, you carry a digit into the next column. If the top digit is smaller than the one below it, you cross out digits to regroup or borrow.

Although this standard algorithm is highly efficient for adults, it can quickly become overwhelming for an anxious upper-elementary child. Keeping track of carried digits, crossed-out numbers, regrouping marks, and long strings of zeroes can place heavy demands on working memory. As the page becomes visually crowded, children often lose their place, become frustrated, and begin guessing simply to finish the problem.

Your focus this week is to move away from one correct way to calculate. Instead, give your child a range of strategies to explore and encourage them to discover which ones make the most sense to them. One child may quickly understand compensation, while another may prefer the clear steps of partial sums. By building a toolkit of flexible strategies, children learn that mathematics is about making sense of numbers, not simply following a set of rules.

□ Science Behind This

When your child encounters traditional column addition or subtraction with heavy regrouping, their brain is forced to perform a delicate juggling act. They have to hold the original values in mind, carry out basic calculations, remember where to place a carried digit, and keep track of which columns they have already completed. It's no surprise that many children lose their place or become overwhelmed.

Flexible strategies like compensation or partial sums reduce that load by breaking the calculation into smaller, easier-to-follow steps. Instead of trying to keep track of everything at once, children can focus on one part of the problem at a time. This frees up working memory, making it easier to think clearly, solve the problem, and notice mistakes.

Continue Your Learning Online

Help your child strengthen their understanding of multi-digit addition and subtraction through visual models and efficient strategies. Focus on understanding the mathematics behind each method while encouraging confidence with increasingly complex problems.

Scan the QR code to open the BridgeKey™ Companion Portal.

BridgeKey™: MB-D4KR

Week 5: Multiplication Structures & Visual Blueprint

Weekly Purpose

This week introduces a major shift in how your child approaches multiplication. Rather than relying only on memorization or one fixed method, your child will learn several strategies that reveal how multiplication works. Area models, partial products, doubling and halving, and breaking numbers into friendlier parts help children solve larger multiplication problems while building understanding, flexibility, and confidence.

MathBridge Reminder

"Understanding comes before memorization".
"Numbers are meant to be explored, not feared".
"Flexible strategies lead to confident thinking".

◈ Parent Learning Focus

If you grew up practicing math through timed multiplication drills, flashcards, or chanting times tables until you were blue in the face, you are not alone. For decades, speed and instant recall were treated as the sign of a good mathematician. But for an anxious child in Grades 3 through 5, nothing triggers a "freeze" response faster than a parent holding up a card and saying, "Quick, what's seven times eight?"

If they don't know the answer instantly, the silence feels heavy. To avoid feeling embarrassed, a child may start guessing random numbers.

Your focus this week is to see visual models as a bridge to understanding, not a crutch. As your child's confidence grows, they will naturally rely on these models less often. There is no need to rush that process. If your child forgets a multiplication fact, the goal is not to panic or guess. The goal is to know what to do next. Encourage your child to draw an area model, break a number into smaller parts, or use another strategy to work out the answer. As children explore different approaches, they begin to see that multiplication is something they can reason through, not simply something they have to remember. Over time, they will discover which strategies make the most sense to them and use them with increasing confidence.

□ Science Behind This

When a child relies only on memorized facts, forgetting just one multiplication fact can quickly trigger panic. Instead of thinking about the next step, they often become stuck because all of their attention shifts to the fact they cannot remember.

Visual models and flexible strategies give children another way forward. Instead of freezing, they can draw an area model, break numbers into smaller parts, or use another strategy to work out the answer. These approaches reduce the pressure on working memory and help children stay engaged with the mathematics. Rather than feeling trapped by what they cannot remember, they learn to reason their way to a solution, one step at a time.

Continue Your Learning Online

Continue developing multiplication skills through visual models, patterns, and meaningful mathematical connections. Encourage your child to explain their thinking, compare strategies, and build understanding that extends beyond memorizing facts.

Scan the QR code to open the BridgeKey™ Companion Portal.

BridgeKey™: MB-X9MF

Week 6: Division Structure & The Choice Toolkit

Weekly Purpose

This week, we make long division manageable. Instead of asking children to memorize one fixed sequence of steps that is easy to lose track of, we introduce two flexible strategies: Partial Quotients and the Reverse Area Model. Whether they divide by subtracting groups or build the quotient one section at a time, children learn to approach long division in a way that makes sense to them. This gives them greater confidence and more control over each step of the calculation. BridgeKey™: MB-M4QD

MathBridge Reminder

"Large problems become easier when broken into smaller steps".
"Understanding each step builds confidence".
"There is more than one path to the correct answer".

◈ Parent Learning Focus

Long division is often one of the most challenging topics children encounter in the upper-elementary years. Many of us learned a fixed sequence of steps: divide, multiply, subtract, bring down, and repeat. Keeping track of each step while estimating, subtracting, and monitoring place value places a heavy demand on working memory. A small mistake early in the calculation can make the rest of the problem difficult to follow, leaving many children feeling frustrated and unsure of themselves.

This week, your child will explore two different ways to approach long division: Partial Quotients and the Reverse Area Model. Rather than memorizing one fixed procedure, they will learn strategies that break large division problems into smaller, more manageable steps. Encourage your child to try both approaches and notice which one feels more natural. As they begin to understand what each step represents, long division becomes less about remembering a sequence and more about making sense of the mathematics.

□ Science Behind This

Traditional long division places a heavy demand on a child's working memory. Children have to estimate, recall multiplication facts, line up numbers correctly, subtract accurately, and keep track of each step. When one part of the process goes wrong, it becomes easy to lose their place, and the entire calculation can quickly feel confusing.

Partial Quotients and the Reverse Area Model reduce this cognitive load by breaking division into smaller, more manageable steps. Instead of trying to find the perfect quotient immediately, children can work with numbers they already understand and develop the solution step by step. This allows them to focus on one step at a time, making it easier to stay organized, recognize patterns, and understand how the answer develops.

Continue Your Learning Online

This week's activities support your child's understanding of division through visual reasoning and flexible problem solving. Allow time to explore different approaches and help your child develop confidence in choosing strategies that make sense to them.

Scan the QR code to open the BridgeKey™ Companion Portal.

BridgeKey™: MB-C7YP

Week 7 Week 7: Multi-Step Word Problems & The Translation Toolkit

Weekly Purpose

This week, we are tackling word problems. Many children feel overwhelmed when they see a page full of text because they are unsure where to begin. Instead of looking for numbers right away, your child will learn to read the story first and focus on understanding what is happening. Once the situation makes sense, they will use simple strategies to organize the important information, identify what the problem is asking, and plan a solution one step at a time. By slowing down and making sense of the story before calculating, word problems become easier to understand and far less intimidating.

MathBridge Reminder

"Understand the story before searching for the numbers".
"Organize the information before solving the problem".
"One step at a time leads to clearer thinking".

◈ Parent Learning Focus

Word problems ask children to do several things at once. They have to read the story, understand what is happening, identify the important information, decide which math operation to use, and carry out the calculation. When children try to manage all of these demands at the same time, they can quickly become overwhelmed. Instead of reading to understand the situation, they often begin searching for keywords such as “in all or “left” and guess which operation to use.

Our main focus this week is to separate understanding the problem from solving it. Many children can complete the mathematics once they understand the problem. The challenge often lies in translating the problem into mathematics. Encourage your child not to reach for a pencil or write an equation until they have first made sense of the story. Help them identify the important details, decide what the problem is asking, and choose an appropriate strategy before they begin calculating.

□ Science Behind This

Imagine trying to follow a new GPS route in a heavy downpour while your favorite music is playing, people in the car are talking, and a police siren is sounding behind you. Before long, it becomes difficult to focus on where you're going.

That is often what solving a word problem feels like for a child. They have to read the story, understand what is happening, identify the important information, remember the numbers, decide which mathematical operation to use. Trying to manage all of these demands at once places a heavy load on working memory, making it much easier to lose track of the problem or guess instead of thinking it through.

Breaking the problem into smaller, manageable steps reduces that cognitive load. When children first focus on understanding the story and then solve one step at a time, they no longer have to keep everything in their minds at once. Recording each step as they go frees up working memory, helping them stay organized, think more clearly, and solve problems with greater confidence.

Continue Your Learning Online

Support your child as they develop confidence with multi-step word problems. Encourage them to slow down, make sense of the situation, organize their thinking, and solve problems one step at a time.

Scan the QR code to open the BridgeKey™ Companion Portal.

BridgeKey™: MB-G2VL

Week 8: Visual Fraction Landscape & The Autonomy Map

Weekly Purpose

This week, we introduce fractions as a way of describing parts of a whole. Rather than relying on memorized rules, your child will use visual strategies to explore sharing, compare the size of fractions, and make sense of how different fractions relate to one another. As their understanding grows, fractions become less abstract and much easier to reason about.

MathBridge Reminder

"Fractions make sense when children can see and discuss the parts of a whole".
"The size of the pieces changes, but the whole remains the same".
"Understanding fraction size is more important than memorizing rules".

◈ Parent Learning Focus

Fractions are often the point where many parents remember mathematics becoming more difficult. For many of us, fractions were introduced as rules and procedures before we fully understood what the numbers represented. Without a strong visual understanding, it is easy for children to become overwhelmed and rely on memorization instead of reasoning.

Your focus this week is to help your child make sense of fractions before asking them to calculate. Encourage them to draw, compare, and talk about fractions using visual models and familiar benchmarks such as 0, 1/2, and 1. If one representation makes more sense to them than another, let them use it. As children begin to see how fractions relate to one another, they develop a stronger understanding and greater confidence before formal procedures are introduced.

□ Science Behind This

When children first encounter fractions, they often apply whole-number thinking. That is why many believe that **1/4** must be larger than **1/2** simply because **4** is greater than **2**. Instead of seeing a fraction as a single number that represents a quantity, they focus on the numerator and denominator as two separate whole numbers. This misunderstanding makes it more difficult to reason about the size and value of fractions.

Visual representations help children build accurate mental models of fractions. Seeing fractions as parts of a whole or as points on a number line helps children connect the symbols to meaningful quantities. Familiar reference points, such as **0**, **1/2**, and **1**, provide stable anchors that make it easier to estimate, compare, and order fractions. As children develop these mental images, they rely less on whole-number thinking and begin to understand fractions as numbers with their own size and value.

Continue Your Learning Online

This week's activities strengthen fraction understanding through visual models and mathematical reasoning. Encourage your child to explore equivalent fractions, compare fractions with confidence, and explain the thinking behind each solution.

Scan the QR code to open the BridgeKey™ Companion Portal.

BridgeKey™: MB-J8RT

Week 9: Mastering Fraction Operations & Mental Workspace Separation

Weekly Purpose

This week helps your child build confidence when adding and subtracting fractions without becoming overwhelmed by the many steps involved. Rather than relying on memorized procedures before understanding, your child will use visual models and learn strategies that keep whole numbers and fractions organized as separate parts of the calculation. They will also explore mixed numbers, learning to see them as whole numbers and fractional parts working together rather than as a completely new type of number.

We are moving from simply recognizing fractions to combining and separating them with understanding. Throughout the week, low-pressure games and hands-on activities provide opportunities to practice these skills in a relaxed, enjoyable way, helping children strengthen their confidence without the pressure of getting every answer right. Success this week is not measured by speed. It is measured by organized thinking, reasoning, and a child who feels confident working through one step at a time.

MathBridge Reminder

"Fractions become easier when whole numbers and fractional parts are organized clearly".
"Understanding each part of the calculation comes before combining them".
"Breaking a complex problem into smaller steps builds confidence and accuracy".

BridgeKey™: MB-B9XF

◈ Parent Learning Focus

Adding and subtracting fractions requires children to coordinate several ideas at once. They must understand that fractions can only be combined when the parts are the same size, keep track of each step, and organize their calculations carefully. Without a clear understanding of why common denominators are needed, the procedure can feel confusing and difficult to remember.

Your focus this week is to help your child understand what is happening before introducing the procedure. Encourage them to use visual models to see how fractions can be divided into equal-sized parts before combining them. When working with mixed numbers, help them keep the whole numbers and fractional parts organized as separate parts of the calculation. An organized approach and logical thinking are more important than finishing quickly.

□ Science Behind This

Keeping whole numbers and fractional parts separate reduces this cognitive load. Visual representations help children see how the same whole can be divided into smaller equal parts, making equivalent fractions easier to understand rather than simply memorize. Together, these strategies help children focus on one step at a time, think more clearly, and solve fraction problems with greater confidence.

Continue Your Learning Online

As you complete the Parent Support Journey, use this week's learning activities to consolidate your child's understanding of fractions while strengthening confidence and independence. Take time to celebrate how far your child has come and continue using the strategies from this book whenever new mathematical challenges arise.

Scan the QR code to open the BridgeKey™ Companion Portal.

BridgeKey™: MB-W6PK

Week 10: Beyond the Bridge – Where Confidence Goes Next

If you have reached this point, take a moment to pause and reflect on how far you and your child have come.

Over the past nine weeks, you have done far more than help your child practice math skills. You have begun rebuilding something much deeper, your child's relationship with mathematics itself.

You have worked to create emotional safety around learning and reduced pressure. You have learned how anxiety affects thinking. You have explored new ways to respond to mistakes, confusion, and frustration. Most importantly, you have helped your child experience mathematics as a place where thinking, persistence, and growth matter more than perfection.

That is no small achievement.

Many parents begin this journey believing that the primary challenge is mathematics. By the time they reach this stage, they often discover that the deeper challenge was confidence, emotional safety, and helping a child believe that they are capable of learning difficult things.

Everything you've done throughout MathBridge has been designed to help your child develop math confidence, independence, and understanding by first building a foundation of deep trust in their own thinking.

I can't stress this enough: when children learn that mistakes can be corrected, that confusion is temporary, and that challenges can be managed without panic or shame, something important begins to change. They become more willing to engage and more willing to take on challenges. They recover more quickly from challenges and become independent learners. That growing independence is one of the most valuable gifts we can offer children.

The Next Stage of the Journey

As mentioned in the early chapters, as children continue through upper primary and beyond, mathematics naturally becomes more demanding and concepts become increasingly connected. Problems require deeper reasoning. Children are asked to explain their thinking, apply multiple skills simultaneously, and navigate unfamiliar situations that do not always have obvious starting points.

As children move into the later years of primary school, fractions become more complex, multi-step problems become more common, and communicating their mathematical thinking becomes more important. Children are expected to compare different strategies and select the most appropriate to solving a problem. They apply several skills simultaneously and eventually begin exploring ideas that prepare them for algebra and more advanced mathematical reasoning. While the mathematics changes, something important remains constant.

Children continue to benefit from the same foundations that they have been building throughout this book:

- Emotional safety
- Conceptual understanding
- Productive struggle
- Independent thinking
- Confidence through experience
- A healthy relationship with challenge

These foundations do not become less important as mathematics becomes more challenging.

They become more important.

A child who trusts their own thinking is far better equipped to navigate complexity than a child who has simply memorized procedures.

A child who understands how to recover from mistakes is far more resilient than a child who believes mistakes should never happen.

When children are less focused on avoiding mistakes, they can devote more of their working memory to mathematical thinking.

This is why the work you have done throughout MathBridge matters.

You have not simply been helping your child learn mathematics. You have been helping them learn how to learn.

What Progress Often Looks Like

As you continue supporting your child through complex math concepts, it is important to remember that progress is not always obvious.

As a parent, you may expect confidence to grow in a straight line. In reality, growth is often uneven. Some weeks your child may appear more confident, independent, and willing to engage. Other weeks they may seem hesitant, frustrated, or uncertain again. This does not mean the progress has disappeared. Confidence is built through a consistent positive learning experiences over time.

Your child may show progress in small ways before you notice bigger changes:

- Perhaps your child starts a task with less resistance.
- Perhaps they recover more quickly after making a mistake.
- Perhaps they ask fewer reassurance questions.
- Perhaps they attempt a difficult problem before asking for help.
- Perhaps they begin explaining their thinking more openly.

These moments may appear small, but they are often powerful indicators that your child's relationship with mathematics is changing. So, I encourage you to celebrate the small wins.

Small wins may not look impressive in the moment, but over time they become the foundation on which lasting confidence is built.

Challenges will always be part of learning mathematics. What changes is how your child responds to them—with greater confidence, resilience, and understanding.

Whenever you find yourself wondering whether progress is happening, try looking beyond marks, speed, and completed worksheets:

- Look for growing independence.
- Look for increased willingness to try.
- Look for healthier responses to mistakes.
- Look for moments of persistence.
- Look for evidence that your child is beginning to trust their own thinking.

Those are often the signs that matter most.

A Final Thought

Children do not need to believe they are naturally gifted at mathematics.

They need to believe they can learn.

They need to believe that mistakes are survivable.

They need to believe that effort can lead to improvement.

They need to believe that confusion is not failure.

They need adults who continue believing in them while they are still learning to believe in themselves.

The confidence your child develops today may influence far more than their future experiences with mathematics.

This is not the end of the bridge.

It is simply the point where your child begins crossing more of it with growing confidence, increasing independence, and a deeper belief in their own ability to learn.

Thank you for allowing MathBridge to be part of your family's learning journey.

MathBridge Glossary

Behavioral Defense Mechanisms

The visible behaviors children use to protect themselves from emotional discomfort during mathematics learning, including avoidance, rushing, excessive help-seeking, perfectionism, humor, and emotional shutdown.

Behavioral Translation Matrix

A framework that helps parents interpret visible behaviors by considering the emotional, cognitive, or psychological experiences occurring beneath the surface.

Cognitive Load

The mental effort required to think, process information, and solve a task.

Cognitive Overload

A state in which the demands of a task exceed a child's available thinking capacity, causing confusion, errors, frustration, or emotional shutdown.

Confidence Before Speed Principle

The principle that understanding, reasoning, and confidence should be prioritized over speed.

Conceptual Understanding

A deep understanding of mathematical relationships, structures, and ideas that allows children to reason flexibly rather than simply memorize procedures.

Developmental Bridge

The gradual transition from concrete experiences and visual models toward increasingly abstract mathematical thinking.

Emotional Overload

A state in which emotional stress interferes with a child's ability to think clearly, remember information, and engage productively with learning.

Emotional Regulation

The ability to manage emotions sufficiently to remain engaged in learning, problem-solving, and recovery from mistakes.

Emotional Safety

A learning condition in which children feel secure enough to make mistakes, ask questions, attempt challenges, and remain engaged during mathematical learning.

Emotional Safety Threshold

The point at which emotional stress begins to override a child's ability to think, remember, and reason effectively.

Gradual Release Of responsibility

A support approach in which adults provide assistance initially and then gradually return responsibility for the thinking process back to the child as confidence and competence increase.

Help Without Taking Over Method

A support approach that allows parents to assist children while preserving the child's ownership of the thinking process.

Hidden Overload Pattern

A pattern in which children experience cognitive and emotional overload that may appear externally as rushing, excessive reassurance seeking, avoidance, perfectionism, or emotional shutdown.

Internal Narrative

The ongoing story children develop about themselves as learners based on their repeated experiences of success, failure, confidence, and struggle.

Learned Helplessness

A pattern in which children lose trust in their ability to think independently and increasingly rely on adults to manage cognitive challenges for them.

Math Anxiety

Feelings of worry or fear about mathematics that make it harder to think clearly and solve problems.

"Looking Beneath" Method

A reflective approach that encourages parents to pause before reacting and consider what emotional or cognitive experience may be driving a child's visible behavior.

Pressure-Avoidance Strategy

A coping strategy in which children avoid, rush, distract themselves, or seek immediate rescue to reduce emotional discomfort.

Productive Struggle

The experience of working through a challenge while remaining able to think, persist, and learn.

Productive Struggle Window

The optimal zone in which children experience sufficient challenge to promote learning while remaining emotionally safe enough to continue thinking effectively.

Rescue Trap

The pattern that develops when adults provide too much assistance too quickly, reducing children's opportunities to develop independence, confidence, and self-trust.

Retrieval System

The brain's ability to access previously learned knowledge and strategies.

Working Memory

The brain's temporary thinking space used to hold and work with information while solving a problem.

www.ingramcontent.com/pod-product-compliance
Lightning Source LLC
LaVergne TN
LVHW021737071126
851857LV00040B/1478
* 9 7 9 8 9 0 5 9 9 8 6 3 8 *